SODIUM DITHIONITE, RONGALITE AND THIOUREA OXIDES

Chemistry and Application

SODIUM DITHIONITE, RONGALITE AND THIOUREA OXIDES

Chemistry and Application

Sergei V. Makarov

Ivanovo State University of Chemistry and Technology, Russia

Attila K. Horváth

University of Pécs, Hungary

Radu Silaghi-Dumitrescu

University of Babes-Bolyai, Romania

Qingyu Gao

University of Mining and Technology, China

NEW JERSEY · LONDON · SINGAPORE · BEIJING · SHANGHAI · HONG KONG · TAIPEI · CHENNAI · TOKYO

Published by

World Scientific Publishing Europe Ltd.

57 Shelton Street, Covent Garden, London WC2H 9HE

Head office: 5 Toh Tuck Link, Singapore 596224

USA office: 27 Warren Street, Suite 401-402, Hackensack, NJ 07601

Library of Congress Cataloging-in-Publication Data
Names: Makarov, Sergei V. (Professor of chemistry)
Title: Sodium dithionite, rongalite, and thiourea oxides : chemistry and application /
    Sergei V. Makarov (Ivanovo State University of Chemistry and Technology, Russia)
    [and three others].
Description: New Jersey : World Scientific, 2016. | Includes bibliographical references.
Identifiers: LCCN 2016014278 | ISBN 9781786340955 (hc : alk. paper)
Subjects: LCSH: Reduction (Chemistry) | Organosulfur compounds. | Sulfur compounds.
Classification: LCC QD63.R4 S63 2016 | DDC 547/.06--dc23
LC record available at https://lccn.loc.gov/2016014278

British Library Cataloguing-in-Publication Data
A catalogue record for this book is available from the British Library.

Desk Editors: Chandrima Maitra/Mary Simpson

Typeset by Stallion Press
Email: enquiries@stallionpress.com

Printed in Singapore

Acknowledgment

This work was supported by Russian Foundation for Basic Research, project 16-03-00162 and Russian Scientific Foundation, agreement No. 14-23-00204 (S. V. M.) as well as by the Hungarian Research Fund (OTKA) Grant No. K116591 (A. K. H.). Q. G. is indebted to the financial support provided by the National Natural Science Foundation of China, Grant No. 21573282 and the Natural Science Foundation of Jiangsu Province, Grant No. BK20131111.

Glossary

AA acrylic acid
ABS acrylonitrile-butadiene-styrene
Adogen methyltrialkyl(C_8–C_{10})ammonium chloride
ADXRD angle-dispersive X-ray diffusion
AIMSA aminoiminomethanesulfinic acid
Aliquat trioctylmethylammonium chloride
APS ammonium persulfate
ATRP atom transfer radical polymerization
BCCW boiler cleaning wastes
B-DNIC binuclear dinitrosyl iron complex
Boc tert-butyl oxycarbonyl
Bz PhCO
CHP cumene hydroperoxide
CL chemiluminescence
CoPc cobalt phthalocyanine
CRGO chemically reduced graphene oxide
CRT cathode ray tube
CSTR continuous-flow stirred tank reactor
CTMP chemithermomechanical pulp
D5 decamethylcyclopentasiloxane
DCHA dicyclohexyl ammonium
DETU N,N'-diethylthiourea
DFT density functional theory
DIBH diisopropylbenzene hydroperoxide
DMF dimethylformamide
DMPO 5,5-dimethylpyrroline-1-oxide
DMSO dimethylsulfoxide
DMTU N,N'-dimethylthiourea
DNIC dinitrosyl iron complex

DPA 2-(diisopropylamino)ethyl methacrylate
DPP differential pulse polarography
DTLRP degenerative chain transfer mediated living radical polymerization
EDTA ethylenediaminetetraacetic acid
EIS electrochemical impedance spectroscopy
ESI-MS electrospray ionization mass spectroscopy
ETC electron transfer cocatalyst
ETU ethylenethiourea
FDS formamidine disulfide
FTIR Fourier transform infrared spectroscopy
GGA generalized gradient approximation
GO graphene oxide
GRS government rubber styrene
HBS sickle hemoglobin
HEC hydroxyethyl cellulose
HMPA hexamethylphosphoramide
HMS hydroxymethanesulfinate
HPLC high performance liquid chromatography
Hr hemerythrin
IC ion chromatography
IR infrared spectroscopy
LDA local density approximation
LRP living radical polymerization
Mb myoglobin
Me_6TREN tris[2-dimethylamino)ethylamine
M-DNIC mononuclear dinitrosyl iron complex
Me methyl
metMb metmyoglobin
MV methylviologen
n-BA n-butyl acrylate
NBO natural bond orbital
nZVI nano zero valent iron
–OSu N(hydroxysuccinimidyl)
OV octylviologen
P porphyrin
P2EHA poly(2-ethylhexyl acrylate)
PAN polyacrylonitrile
PBA poly(n-butyl acrylate)
PCB polychlorinated biphenyl

PEG polyethylene glycol
Ph phenyl
Pht o-OCC$_6$H$_4$CO$^-$
PS polysterene
PTC phase-transfer catalysis
PVC polyvinyl chloride
PVP polyvinylpyrrolidone
RAFT reversible addition-fragmentation chain transfer
RGH reduced graphene hydrogel
RGO reduced graphene oxide
SET single electron transfer
SOM soft-oxometalate
Tacn triazacyclononane
TBAB tetrabutylammonium bromide
TBHP tert-butyl hydroperoxide
TCE trichloroethylene
TDAE tetrakis(dimethylamino)ethylene
TDO thiourea dioxide
THF tetrahydrofuran
THPS tetrakis(hydroxymethyl)phosphonium sulfate
TLC thin-layer chromatography
TMEDA tetramethyl ethylene diamine
TMO thiourea monoxide
TMP thermomechanical pulp
TSPc tetrasulfophthalocyanine
TTO thiourea trioxide
TU thiourea
VC vinyl chloride
XPS X-ray photoelectron spectroscopy

Contents

Acknowledgment   v

Glossary   vii

List of Figures   xiii

List of Table   xix

About the Authors   xxi

1. Introduction   1

2. Synthesis   5

    2.1   Dithionites   5
    2.2   Sodium Hydroxymethanesulfinate (Rongalite)
        and Its Relatives   9
    2.3   Thiourea Oxides   11

3. Structure   21

    3.1   Dithionites   21
    3.2   Hydroxymethanesulfinates   26
    3.3   Thiourea Oxides   27

4. General Properties and Analysis   33

5. Stability in Solutions under Anaerobic and Aerobic Conditions   39

6. Organic Reactions   53

    6.1   Synthesis of Organofluorine Compounds   53

6.2   Synthesis of Guanidines   59
6.3   Synthesis of Sulfur-, Selenium- and Tellurium-Containing
       Compounds   66
6.4   Reduction of Aldehydes, Ketones
       and Unsaturated Compounds   80
6.5   Synthesis of Nitrogen-Containing Compounds   87
6.6   Organocatalytic Reactions   96

7. Inorganic Reactions and Material Chemistry   107

7.1   Reduction of Graphene and Graphite Oxides   107
7.2   Reduction of Metal Complexes.
       Synthesis of Metals and Metal Sulfides   115
7.3   Reduction of Halogen Compounds   129

8. Industrial Applications   139

8.1   Textile Industry   139
8.2   Paper Industry   143
8.3   Polymerization Processes   147

9. Miscellaneous   157

Bibliography   159

Index   209

List of Figures

2.1 Proposed mechanism of oxidation of thiourea by hydrogen peroxide 16

2.2 Proposed mechanism for the Ru^{III}(edta)-catalyzed oxidation of TU by HSO_5^- 16

2.3 Oxidation of ETU to the corresponding sulfenic (n = 1), sulfinic (n = 2) and sulfonic (n = 3) acids 19

3.1 Observed structure of dithionate (a), $S_2O_5^{2-}$ (b), predicted structure of dithionite (c) and dithionite (d) 22

3.2 ORTEP projection of $[Cp^*Mo(CO)_3]_2$ $(\mu\text{-}S_2O_4)$ 24

3.3 Reversible photochromic transformation of $[(RhCp^*)_2(\mu\text{-}CH_2)_2(\mu\text{-}O_2SSO_2)]$ (*: asymmetric sulfur atom) 24

3.4 Structure of complexes 3 and 4 25

3.5 Structure of complex (5) 26

3.6 Structure of complex (6). Hydrogen atoms are omitted for clarity 26

3.7 Possible structures of hydroxymethanesulfinate in aqueous solution 27

3.8 Structure of zinc hydroxymethanesulfinate 28

3.9 Polymer of zinc hydroxymethanesulfinate 28

3.10 Redox isomers of TDO 29

3.11 Carbenoid structure of TDO 29

5.1 Possible reaction pathways for O_2 release from TDO 51

6.1 Radical mechanism of sulfinatodehalogenation reaction 55

6.2 Synthesis of fluorinated extended porphyrins using modified sulfinatodehalogenation reaction 56

6.3   Reaction of 4-acetylpyridine with sodium
      trifluoromethanesulfinate under oxidative conditions
      (catalyst – $FeSO_4$, $CuSO_4$ or $CoClO_4$)   57
6.4   Reaction of perfluoroalkyl iodides with coumarins   58
6.5   Synthesis of 3-perfluoroalkylated coumarins (X = O),
      thiocoumarins (X = S) and 2-quinolones (X = NR) by direct
      perfluoroalkylation with perfluoroalkyl iodides and sodium
      hydroxymethanesulfinate   58
6.6   L-Arginine   60
6.7   Reaction of thiourea trioxide with amine   61
6.8   Reaction of thiourea trioxide with methyl anthranilates   63
6.9   Guanidines synthesized from thiourea trioxide   63
6.10 DuP 714 and a guanidino-containing thrombin inhibitor   64
6.11 A guanidino-containing thrombin inhibitor synthesized
      by Lu *et al.*   64
6.12 Reaction of thiourea trioxide with chitosan   65
6.13 Synthesis of β-β'-disubstituted diethyl sulfones   67
6.14 Reaction of 1,4-benzoquinone with hydroxymethanesulfinate . .   67
6.15 Synthesis of symmetrical sulfones $R-CH_2-SO_2-CH_2-R$
      from sodium hydroxymethanesulfinate and Mannich bases . . .   68
6.16 Synthesis of cyclic sulfone   68
6.17 Synthesis of dibenzyl sulfones   69
6.18 Synthesis of sulfone from α, α'-dibromo-o-xylene   70
6.19 Synthesis of benzosulfones via rongalite in the presence of TBAB   70
6.20 Synthesis of olefinic sulfones from alkenyl bromides
      and rongalite   70
6.21 Proposed mechanism for reaction of alkenyl bromides
      and rongalite   71
6.22 Classical Julia olefination   72
6.23 Synthesis of olefins from vinylsulfones   72
6.24 Cyclic pathway for the viologen-mediated reductive
      desulfonylation of α-nitro-sulfones by sodium dithionite
      (V^{2+}–viologens (1,1'-dialky1-4,4'-bipyridiniums))   73
6.25 Proposed mechanism for the viologen-mediated desulfonylation
      of α-nitro sulfones   73
6.26 Synthesis of sultines   74
6.27 Structure of 1,4-thiatellurane and tellurophene   75
6.28 Selective formation of allyl alcohols   75
6.29 Parallel formation of tellurophenes and allyl alcohols   76

6.30 Synthesis of tellurophene and allylic alcohol 76

6.31 Synthesis of tetrahydroselenophene
and tetrahydrotellurophene 77

6.32 Proposed mechanism of the synthesis of allylic alcohols 77

6.33 Synthesis of β-hydroxy sulfides 78

6.34 Proposed mechanism of synthesis of β-hydroxy sulfides. 78

6.35 Synthesis of β-sulfido carbonyl compounds 78

6.36 Stereoselective protocol for hydrothiolation of terminal
alkynes 79

6.37 Possible mechanism of synthesis of (Z)-1-alkenyl sulfides
(selenides) 79

6.38 Synthesis of thioesters and selenoesters 79

6.39 Synthesis of fluorous seleninic acid 79

6.40 By-products of reduction of benzils and 4-chlorophenacyl
bromide 84

6.41 Reaction of phenacyl chloride and sodium
hydroxymethanesulfinate 84

6.42 Deoxygenation of α,β-epoxy ketones 85

6.43 Chemoselective reduction of aldehydes using thiourea dioxide . 85

6.44 Chemoselective reduction of aromatic nitrocarbonyls to the
corresponding nitroalcohols 85

6.45 Reduction of N-substituted noroxymorphone derivatives
by TDO 87

6.46 Structure of different N-heterocycles synthesized by TDO . . . 88

6.47 Reaction of thiourea dioxide in ethanolic alkali with
2,2'-dinitrobiphenyl and several related dinitro compounds . . . 89

6.48 Synthesis of pyrroles 90

6.49 Diones (**8**) and products of their reactions with TDO (**9,10**) . 90

6.50 Thiourea trioxides used in reactions with azides 91

6.51 Nitrogen-containing heterocycles produced from thiourea
trioxides 91

6.52 Deoxygenation of N-oxides 91

6.53 Synthesis of 1,2-disubstituted benzimidazoles 92

6.54 Reduction of methyl viologen 93

6.55 Coupled redox reaction NAD$^+$/NADH 95

6.56 Possible structures of sulfinate adduct 95

6.57 Possible mechanism of the formation of NADH analogs 96

6.58 Biginelli condensation using PEG–TDO complex 98

6.59 Plausible mechanism for Biginelli condensation 99

6.60 Synthesis of dihydropyrido[2,3-d]pyrimidine-2,4-dione 99
6.61 TDO catalyzed one-pot synthesis of heterocycles 100
6.62 Probable mechanism of synthesis of coumarin 101
6.63 Synthesis of naphthopyran 101
6.64 Plausible mechanism for the TDO-catalyzed synthesis
       of naphthopyrans 102
6.65 Synthesis of 1,8-dioxo-octahydroxanthenes 102
6.66 Synthesis of polyhydroquinolines 103
6.67 TDO-catalyzed oxidation of sulfides to sulfoxides
       with TBHP 103
6.68 Plausible oxidation pathway of oxidation of sulfides to sulfoxides
       in presence of TDO 103
6.69 TDO-catalyzed oxidation of alcohols with TBHP 104
6.70 Combined metal and organocatalyst promoted hydrolysis
       of imines 104
6.71 Probable mechanism of catalytic hydrolysis of imines 105

7.1   Lerf–Klinowski model of graphite oxide with the omission
       of minor groups (carboxyl, carbonyl, ester, etc.) on the periphery
       of the carbon plane of the graphitic platelets of graphite oxide 108
7.2   Synthesis of graphene from graphite 109
7.3   Functional groups on GO reduced by TDO 112
7.4   Proposed mechanism for the reduction of GO by TDO in alkaline
       aqueous solution 114
7.5   Possible consecutive pathway for reduction
       of methemerythrin 119
7.6   Mechanism of reaction between heme and trichloromethanes
       in the presence of dithionite 120
7.7   Structure of bis(hydroxylamino)hexaaza cryptand 128

8.1   Reaction of indigo dye to leucoform (soluble form) 140
8.2   Reaction of anthraquinone derivatives with dithionite
       or α-hydroxyethanesulfinate 140
8.3   Reaction of anthraquinone derivatives with sulfoxylate 141
8.4   Reaction of dithionite with indigocarmine 142
8.5   Reaction of dithionite with lignin chromophores 144
8.6   Reduction of coniferaldehyde-type structures during dithionite
       bleaching 144
8.7   Structure of compound 1 146

8.8   Reactions in the system peroxodisulfate, ferrous sulfate, EDTA
and HMS 148
8.9   Initiation with CHI_3 and mediated by $Na_2S_2O_4$ SET
reactions 151
8.10 SET step mediated by ETC 152

List of Table

3.1 Selected bond distances (Å) of some thiourea oxides 29

About the Authors

Attila K. Horváth is an Associate Professor of Chemistry at the Department of Inorganic Chemistry, University of Pécs, Hungary since 2011. He received his MSc (1994) in Chemistry and in Mathematics and a PhD (2000) in Chemistry from University of Szeged, Hungary. He spent two years as a postdoc at the Chemistry Department, Brandeis University, Waltham, MA, USA and he has also been a visiting scholar at the School of Chemical Engineering, China University of Mining and Technology since 2010. His current research interest mainly focuses on elucidating the kinetics and mechanism of the reactions between sulfur- and halogen-containing inorganic species, including those ones that often exhibit nonlinear dynamic phenomena. He has published 56 scientific papers in prestigous (mainly Q1) journals.

Qingyu Gao, born in 1965, received his doctorate in Physical Chemistry from Nankai University in 1996. He has worked as a postdoctoral fellow and visiting scholar at several universities, such as Stanford University, University of Missouri-Columbia, University of Windsor, Brandeis University and Boston University. In 1998, he received a faculty position in China University of Mining and Technology and was promoted to full Professor in 2002. His research interests include spatiotemporal dynamics and active matter, sulfur chemistry related to complex reactions and clean energy. He has published more than 100 papers in the above research fields.

Sergei V. Makarov received his PhD from Ivanovo State University of Chemistry and Technology (ISUCT, Russia) in 1986. In 2003, he was appointed as a Professor of Physical Chemistry at ISUCT, where he is currently the Head of Department of Food Chemistry. He has worked as a visiting scholar at West Virginia University (USA), University of Erlangen-Nuremberg (Germany) and China University of Mining and Technology (China). He is a member of the Editorial Board for *Journal of Coordination Chemistry*. His research interests include chemistry of sulfur-containing reducing agents, kinetics and mechanisms of reactions catalyzed by metallophthalocyanines, and redox transformations of cobalamins.

Radu Silaghi-Dumitrescu holds a PhD from the University of Georgia in Athens, GA, USA (2004) and one from the "Babeş-Bolyai" University in Cluj-Napoca, Romania (2005). He was a Senior Research Officer at the University of Essex, UK, (2004–2006) working on the characterization of hemoglobin-based blood substitutes in the EU-funded "Eurobloodsubstitutes" consortium. Since 2007, he is an Associate Professor at the "Babeş-Bolyai" University, where he also became a Co-Director of the Institute of Technology and a Co-Director of the Center for Molecular Modeling and Computational Quantum Chemistry in 2011, and President of the Scientific Council of the University since 2012. He was a member of the Chemistry and Biochemistry/Biology Commissions of the Romanian National Council for Scientific Research and of the National Council for Accreditation of University Degrees and Titles, and is currently Editor-in-Chief of the newly-founded journal *Acta Metallomica*, edited by the Romanian Academy. His current research interests are centered on small molecule activation by metalloproteins with focus on unusual oxidation states and with ramifications into oxidative and nitrosative stress, blood substitutes, anticancer drugs, natural extracts with antioxidant and biological effects, and biopolymer structure. The tools currently employed for this purpose include computational chemistry, UV–vis, EPR, vibrational and NMR spectroscopy, mass spectrometry, stopped-flow, rapid freeze-quench, protein cloning, overexpression and purification, protein chemical derivatization, animal models, tissue/cell studies.

Chapter 1

Introduction

Sodium dithionite ($Na_2S_2O_4$), sodium hydroxymethanesulfinate (rongalite) ($HOCH_2SO_2Na$) (HMS) and thiourea dioxide ($(NH_2)_2CSO_2$) (TDO) have long been used in chemistry and chemical technology as reductants [1]. Recently, new fields of their application have been developed: reduction of graphene [2–8] and graphite oxides [9], synthesis of metal sulfides [10,11] and nanometer metal powders [12], metallization of fibers [13], organic synthesis [1,14,15] including organocatalytic reactions [16–22] and finally nonlinear phenomena in chemical kinetics [1]. Many newly discovered "relatives" of these compounds have been synthesized [1]. In the past few years, new and important information has been published on dithionite, particularly on the synthesis of complexes with uranium and f-elements as well as with decamethylsamarocene [23–25]. Other notable examples include the studies on photoisomerization of photochromic dithionite complexes in chiral crystals [26–30] and synthesis of the first polythionite [31]. It is found that to receive stable polythionite salt, a countercation should be sufficiently large. The same conclusion has been drawn for sulfoxylates [32]. It is shown that the most convenient precursor for synthesis of sulfoxylates in aqueous solution is thiourea dioxide.

The year 1870 may be considered as a starting point of the development of the chemistry of the above-mentioned sulfur-containing reducing agents. In that year, Schützenberger on the basis of Schönbein's early observations [33] successfully prepared sodium dithionite for the first time [34,35]. To obtain even more stable reductants, later, in the end of 19th century as well as in the beginning of 20th century, a few α-hydroxyalkanesulfinates (including the most important sodium hydroxymethanesulfinate) were synthesized [36]. In 1910, TDO was prepared from the direct reaction of thiourea and hydrogen peroxide [37].

The careers of many eminent chemists were closely related to these chemicals. As a demonstrative example, the topic of the thesis of Nobel laureate Ernst Otto Fischer was "Mechanisms of reactions of carbon monoxide with nickel salts in the presence of dithionites and sulfoxylates" [38]. This thesis, under the direction of the famous German scientist Walter Hieber, a pioneer in the study of metal carbonyls, has become a significant contribution to the chemistry of sulfur-containing reducing agents. Later in this book, we shall also meet other famous chemists involved in the chemistry of these compounds.

Though using dithionite, hydroxymethanesulfinate and TDO as strong reducing agents remains the main field of their application, other important trends should also be highlighted. For example, in contrast to the redox reactions in which the sulfur-containing part of TDO plays the governing role, it is the nitrogen-containing part of TDO and trioxide (TTO) that determines their reactivity in the synthesis of guanidines and its derivatives [39]. The other important field is the synthesis of sulfones, where sulfur-oxygen fragments from dithionite, hydroxymethanesulfinate and TDO are embedded in the target compound [40].

And, of course, TDOs and TTOs deserve a special attention as they can be considered intermediates of the oxidation of thiourea and that of substituted thioureas in some processes. Indeed, the role and significance of thioureas have increased dramatically in the last years as a result of their versatile usage in supramolecular chemistry [41] and organocatalysis [42]. Different thioureas are now among the main environmentally friendly organocatalysts. The existence of two or three oxygen atoms capable of forming additional hydrogen bonds can improve catalytic properties of thiourea oxides [16–22]. However, in contrast to thiourea-catalyzed reactions, the mechanistic features of TDO-catalyzed reactions have been much less studied. TTO has not been used as an organocatalyst at all.

Unfortunately, there is only one relatively old review devoted to all these important sulfur-containing reductants with C–S and S–S bonds: dithionite, rongalite (hydroxymethanesulfinate) and TDO [1]. Since then, one short review on these compounds was published in 2013 (dithionite [43]), one in 2012 (rongalite in organic synthesis [14]), and an additional one about the product of their decomposition sulfoxylic acid [44]. In 2014, we also published a minireview on thiourea oxides [45]. It should be noted that the above-mentioned reviews, besides review on the application of rongalite in organic synthesis, are relatively short ones. There is no comprehensive and comparative review or book on the chemistry and application of these compounds. We believe that there is room for such a book and certainly hope that it will be interesting for a wide audience, especially for specialists in sulfur chemistry, coordination chemistry, biochemistry, nonlinear chemical kinetics, researchers of graphene and graphite, synthetic organic chemists, specialists in applied chemistry (paper, textile, polymer industries) and even specialists in building materials [46–48] and in history of chemistry.

Chapter 2

Synthesis

2.1 Dithionites

Among all sulfur-containing reductants discussed in this book, sodium dithionite was synthesized first. In 1854, a famous chemist, the author of many unexpected observations [49,50], Christian Friedrich Schönbein reported that an aqueous solution of SO_2 in contact with zinc turned rapidly yellow, meanwhile it was also capable of decolorizing a solution of indigo and litmus [33]. After a short period of time the solution deposited sulfur and lost its activity. Later, in 1860s French physician and chemist Paul Schützenberger (in a paper [51] authored by de Vries *et al.*, his initials were falsely indicated as M. P. Since this paper was written in French, M thus stands for Monsieur as J. Wisniak (Ben-Gurion University) kindly paid our attention to this mistake) tried to isolate an active compound, but his experiments with SO_2 and zinc were unsuccessful because the decolorizing power was lost very rapidly [35]. Better results were obtained when the SO_2 (sulfurous acid) solution was replaced by concentrated solution of sodium bisulfite. In this case the reducing power was stronger and lasted for substantially more time, if the solution was kept out of contact with air. Schützenberger named the active substance sodium hydrosulfite and assigned an erroneous formula $NaHSO_2 \cdot H_2O$ [51]. Anyway, discovery of the strong reductant now having named sodium dithionite was among Schützenberger's most valuable achievements and is mentioned in all related textbooks [49].

BASF was the first company to produce sodium dithionite in powder form, in 1906, by zinc dust process [52]. Zinc reacted with sulfur dioxide (bisulfite) in aqueous solution to produce zinc dithionite, which was transformed to sodium dithionite by adding sodium hydroxide. The main reactions are as follows:

$$Zn + 2SO_2 \longrightarrow ZnS_2O_4, \qquad (2.1)$$

$$ZnS_2O_4 + 2NaOH \longrightarrow Na_2S_2O_4 + Zn(OH)_2. \qquad (2.2)$$

The product obtained was named Hydrosulfite (probably after Schützenberger) Conc. BASF [53]. The name "hydrosulfite" conveys the mistaken notion that the compound contains hydrogen. By the time this mistake was recognized, the brand name "Hydrosulfite Conc. BASF" had become so well-known commercially that the company decided not to change the name.

Later on, a process using sodium amalgam was developed. After that BASF elaborated the process in which sodium formate reduces bisulfite to sodium dithionite [54]. For reduction of bisulfite one may also use sodium borohydride [52]. Nowadays dithionite synthesis is performed mainly by formate method. Using this procedure anyone can receive a product containing about 88% of sodium dithionite. In this process [54] sodium formate, dissolved in 80% aqueous methanol, is driven into a stirred vessel. At a pressure of 2–3 bar, sulfur dioxide and sodium hydroxide are introduced into this solution in such a way that pH of 4–5 is maintained. The reaction can be described by the following equation:

$$HCOONa + 2SO_2 + NaOH \longrightarrow Na_2S_2O_4 + CO_2 + H_2O. \qquad (2.3)$$

Under the conditions mentioned above, anhydrous sodium dithionite precipitates as fine crystals. It is filtered, washed by methanol and dried. The main by-products are sulfite, sulfate, thiosulfate, chloride, hydroxyethanesulfonate, hydroxyethanethiosulfate [55], thiodiglycol and 2,2'-dithiodiethanol [56]. Last two are offered to be used as the components of lubricants [56].

A mechanism of reaction between sulfur dioxide and formic acid (formate) was studied by Goliath and Lindgren [57]. This reaction was found to obey a second-order kinetics. The rate is maximum at pH = 2–3. The following reaction pathway was suggested:

$$HCOO^- + SO_2 \longrightarrow {}^\bullet CO_2^- + {}^\bullet SO_2^- + H^+, \qquad (2.4)$$

$${}^\bullet CO_2^- + SO_2 \longrightarrow CO_2 + {}^\bullet SO_2^-, \qquad (2.5)$$

$$2 {}^\bullet SO_2^- \rightleftharpoons S_2O_4^{2-}. \qquad (2.6)$$

As mentioned above, commercial sodium dithionite is not available in its pure form. McKenna and coworkers [58] have suggested a convenient, detailed procedure for the recrystallization of commercial dithionite from

0.1 M NaOH–methanol under anaerobic conditions. Twice-recrystallized dithionite had a purity of $99\pm1\%$ by UV spectroscopy (A_{315}) and elemental analysis. The influence of dithionite quality on the apparent reduction activities of the nitrogenase components (Av1 and Av2) from *Azotobacter vinelandii* was investigated. A significant underestimation of Av2 was shown to occur when deteriorated dithionite was used due to the fact that the $S_2O_4^{2-}/HSO_3^-$ ratio sets the redox potential of the nitrogenase assay solutions (it is known that the main impurity in the dithionite samples is sulfite [58]). It should be noted that, to the best of our knowledge, this is the only work where 99% sodium dithionite was used. This shows that the quality of commercial dithionite (85–88%) is satisfied for the majority of goals.

Lithium dithionite $Li_2S_2O_4$ with a purity of 90% can be also obtained by "formate method" by the reduction of SO_2 using aqueous lithium formate as the reducing agent [59]. Sodium, lithium and potassium dithionites can also be synthesized by the reaction of sulfur dioxide by metal tri-sec-butylborohydride in tetrahydrofuran (THF) at low temperature ($-78°C$) [60]:

$$2MB(CH(CH_3)C_2H_5)H + 2SO_2 + 2THF \longrightarrow M_2S_2O_4$$
$$+ 2THF : B(CH(CH_3)C_2H_5) + H_2. \qquad (2.7)$$

The reaction yields lithium and sodium dithionites with purities 75–88% and potassium salt with purity less than 70%. It was shown that properties of dithionites obtained from aqueous and non-aqueous solutions are significantly different (see later).

Tin(II)-dithionite was first prepared in the impure microcrystalline state by Brunck [61]. In the pure form $Sn_2(S_2O_4)_2$ has been synthesized by reaction of tin powder with liquid sulfur dioxide in presence of water [62]. After one week, crystals of tin(II)-dithionite are formed. In the absence of water no reaction takes place at all.

In attempts to receive organosoluble dithionite (sodium dithionite is almost insoluble in most of the organic solvents), Lough and McDonald have synthesized tetraethylammonium dithionite $(NEt_4)_2S_2O_4$ using anion exchange chromatography on Bio-Rex 5 anion-exchange resins under strictly anaerobic conditions [63]. In 1978, Mincey and Traylor published a method of synthesis of solid 18-Crown-6 sodium salt of dithionite [64]. Recently, however, Bruna and coworkers have shown that [18-Crown-6 Na]$_2S_2O_4$ complex can be prepared in methanol solution but it dissociates into 18-Crown-6 (s) and $Na_2S_2O_4$ (s) on removal of the solvent [65]. Their calculations show that complexation of alkali metal dianion salts to crown ethers are much less favorable than that of the corresponding monoanion salts in the solid state and that the formation of alkali metal crown complexes of stable simple oxy-dianion salts is unlikely. Two years later, the same group showed that gaseous SO_2 reacts with tetrakis(dimethylamino)-ethylene (TDAE) in acetonitrile in a 2:1 stoichiometric ratio to give analytically pure insoluble purple $(TDAE)(O_2SSO_2)$ [31]. They also obtained crystals of $(TDAE)(O_2SSSSO_2)$ from orange solution over the purple solid. This compound is the first example of polythionite.

Thermodynamic estimates show $(TDAE)(O_2SSSSO_2)$ to be stable with respect to loss of sulfur and formation of $(TDAE)(O_2SSO_2)$, in contrast to $[O_2SSSSO_2]^{2-}$ salts of small cations that are unstable toward the related dissociation. Authors concluded that with usage of sufficiently large countercations the chemistry of sulfur oxyanions can be vastly extended.

It should be noted that some controversy exists in the literature on the term "polythionite". Thus, Potteau *et al.* name "polythionite" solutions $Li(SO_2)_n$–HMPA (hexamethylphosphoramide, HMPA) where SO_2 is reduced chemically by lithium [66]. The overall reaction for the preparation of polythionite solutions is the following:

$$Li + nSO_2 \longrightarrow Li^+ + SO_2^- + (n-1)SO_2. \quad \text{with } n \geq 1. \qquad (2.8)$$

Therefore, considering the term "polythionite", one can see two different approaches. One, taken by Bruna *et al.* [31], name compounds by analogy with polysulfides and polythionates, where number of sulfur atoms varies. In the other [66], polythionite has more than one reduced sulfur dioxide species ("monothionite").

Methods of synthesis of dithionite transition-metal complexes can readily be divided into two groups. In the first one, reactions of metathesis are used. Thus, complexes $[(Rh(C_5H_4R))_2(\mu-CH_2)_2(\mu-O_2SSO_2)_2]$ (R = Me, Ph, Et, nPr, CH_2Ph), in which dithionite ligand is side-on coordinated by both sulfur atoms, were prepared by salt metathesis of sodium dithionite with the corresponding rhodium chloride precursors [26–29,67,68]. In the second one SO_2 serves as a precursor. For example, complex $[((\eta^5-C_5Me_5)Mo(CO)_3)_2(S_2O_4)]$, in which dithionite ligand also coordinates with the sulfur atoms to the metal centers, was obtained by a reduction of SO_2 with $[(C_5Me_5)Mo(CO)_3H]$ [69]. Since large number of d- and f-metal complexes can be employed as reducing agents [24], the number of dithionite complexes can be significantly extended.

Indeed, the reaction of decamethylytterbocene $[(\eta^{5-}C_5Me_5)_2Yb$ (THF)$_2]$ with SO$_2$ at low temperature gave two new compounds, namely, the Yb^{III} dithionite/sulfinate complex $[(\eta^{5-}C_5Me_5)_2Yb(\mu^3 1\,k^2O^{1,3},^2k^3$ $O^{2,2,4}$-S$_2$O$_4)_2$ ((η^5-C$_5$Me$_5$) Yb(μ,1kO,2kO-C$_5$Me$_5$SO$_2$))$_2]$ and the Yb^{III} dithionite complex $[((\eta^5$-C$_5$Me$_5)_2$Yb)$_2(\mu$,1k^{2}O^{1,3},2k^{2}O^{2,4}-S$_2$O$_4)]$. These compounds are the first dithionite and sulfinate complexes of f-elements [24].

The first dithionite $[((^{nP,Me}ArO)_3$tacn)U^{IV}2(μ-k^2:k^2-S$_2$O$_4)]$ (tacn = tri-azacyclononane) complex of uranium together with the correspond-ing sulfite complex from reaction of gaseous SO$_2$ with trivalent $[((^{nP,Me}ArO)_3$tacn)$_3$U$^{III}]$ has been prepared by the same group [23].

The reaction of $[(\eta^5$-C$_5$Me$_5)_2$Sm(THF)$_2]$ (tetrahydrofuran, THF) with SO$_2$ resulted in four different products including the dimeric samarium complex $[(\eta^5$-C$_5$Me$_5)_2$Sm(C$_5$Me$_5$SO$_2)_2]$, the dithionite–sulfinate complex $[(\eta^5$-C$_5$Me$_5)_2$Sm(S$_2$O$_4)_2(\eta^5$-C$_5$Me$_5$)Sm(C$_5$Me$_5$SO$_2)_2]$, the dithionite com-plex $[(\eta^5$-C$_5$Me$_5)_2$Sm$_2$(S$_2$O$_4)]$ and the sulfinate complex $[(\eta^5$-C$_5$Me$_5$)Sm$_2$ (C$_5$Me$_5$SO$_2)_4]$. As major reaction pathways, the reductive coupling of two SO$_2$ molecules to form dithionite anion S$_2$O$_4^2$ and nucleophilic attack of one samarocene C$_5$Me$_5$ ligand on the sulfur atom of SO$_2$ were observed [25].

2.2 Sodium Hydroxymethanesulfinate (Rongalite) and Its Relatives

The instability of dithionite in aqueous solution, especially under acidic conditions is well known. Therefore, after its synthesis a hunt for a more stable analog has been fueled almost immediately. As a result, the first successful attempt was the synthesis of α-hydroxyalkanesulfinates, combin-ing high reducing power with stability in aqueous solutions [43]. The most famous member of this group of compounds is sodium hydroxymethane-sulfinate (HMS) or formaldehyde sulfoxylate HOCH$_2$SO$_2$Na. Its technical name is rongalite (in French "rongeage" means discharge). The sodium salt of a hydroxymethanesulfinic acid was first prepared in 1905 [70].

HMS is a product of the reaction of formaldehyde with dithionite [71] (together with hydroxymethanesulfonate HOCH$_2$SO$_3$Na) or the reduction of a mixture of bisulfite and formaldehyde with zinc [72]:

$$S_2O_4^{2-} + 2CH_2O \longrightarrow HOCH_2SO_2^- + HOCH_2SO_3^-, \qquad (2.9)$$

$$HSO_3^- + CH_2O + Zn + H_2O \longrightarrow HOCH_2SO_2^- + Zn(OH)_2. \qquad (2.10)$$

A variant of the second process involves the reduction of aqueous sulfur dioxide with zinc dust to zinc dithionite (see Eq. (2.1)), which is then converted with formaldehyde into the zinc salts of hydroxymethanesulfonic and hydroxymethanesulfinic acids. The sulfonate is reduced to sulfinate with zinc in the presence of sodium hydroxide [36]:

$$ZnS_2O_4 + 2H_2O + 4CH_2O \longrightarrow (HOCH_2SO_3)_2Zn + (HOCH_2SO_2)_2Zn,$$

$$(2.11)$$

$$(HOCH_2SO_3)_2Zn + 2Zn \longrightarrow 2(-OCH_2SO_2)Zn + ZnO + H_2O, \quad (2.12)$$

$$(-OCH_2SO_2)Zn \overset{NaOH}{\longrightarrow} HOCH_2SO_2Na + ZnO. \quad (2.13)$$

The synthesis also produces water-soluble zinc HMS $(HOCH_2SO_2)_2Zn$ together with the corresponding sulfonate $(HOCH_2SO_3)_2Zn$ [36]. The former one can be separated by evaporation *in vacuo* and subsequent drying. The sparingly soluble zinc oxidomethanesulfinate $(OCH_2SO_2)Zn$ can be obtained from zinc HMS by addition of zinc salts and sodium hydroxide. Calcium salt is precipitated by treating sodium HMS with $CaCl_2$. Zinc and calcium HMSs have commercial names Decrolin and Rongalite H, respectively [36].

Mulliez and Naudy have synthesized various relatives of HMS $HOCHRSO_2Na$ (R = H, Ph, Me, p$-HOC_6H_4$, p$-CF_3C_6H_4$, CF_3) and $HOCHCF_3H_2^+N(C_6H_{11})_2$ by slowly adding aldehydes to the mixture of sodium dithionite and sodium hydroxide [71]. Contrary to sulfinates, the corresponding sulfonates are unstable in alkaline solutions and produce aldehydes and sulfite which can be separated from α-hydroxyalkanesulfinates due to their different solubility.

Sodium α-hydroxyethanesulfinate $CH_3CH(OH)SO_2Na$ was prepared by Nooi and coworkers using reaction of photoexcited SO_2 and ethanol at low temperatures with further neutralization with aqueous $NaHCO_3$ solution [73]. Zinc α-oxidoethanesulfinate have been described as well [36].

On the basis of α-hydroxyethanesulfinate BASF has developed reductant Rongal A $N[(CH_3)CHSO_2Na]_3$, obtained in a stable powder form [74]. Aminoderivatives of α-hydroxyalkanesulfinates are prepared by their reactions with ammonia or amines [75]. Thus, Makarov and coworkers have obtained sodium dimethyl- and diethylaminomethanesulfinates [75]. N-methane- or p-toluene-sulfonyl α-aminosulfinic salts were synthesized in a Mannich-like reaction by coupling α-hydroxysulfinates with methane- or toluene-sulfonamides in an aqueous basic medium [76]. Variously N-protected α-aminomethanesulfinates $YNHCH_2SO_2Na$ (Y = protecting group) have been synthesized by coupling of α-aminomethanesulfinate

$H_2NCH_2SO_2Na$ with acylating reagents YX in H_2O–dioxane solutions at room temperature [77]. The dicyclohexylammonium (DCHA) salts $YNHCH_2SO_2H_2N(C_6H_{11})_2$ were prepared by the reaction of corresponding sodium salt with citric acid and dicyclohexylamine (i.e. tris-dicyclohexylammonium citrate). Acylating reagents were taken as Z-benzotriazolyl, –OSu, Boc–O–Boc, Pht–NCOOEt, Bz–imidazolyl, where $Z = PhCH_2OCO$, –OSu = N-(hydroxysuccinimidyl), Boc = Me_3COCO, Pht = o–$OCC_6H_4CO^-$, Bz = PhCO. Yields of Na and DCHA salts are 50–100% and 17–80%, respectively.

Owing to easy release of human carcinogen formaldehyde from hydroxymethanesulfinate, there is an urgent need to develop an ecological alternative process [78,79]. Recently the formaldehyde-free reducing agent Bruggolite® FF6 — disodium α-hydroxyethanoicsulfinate or disodium glyoxylate sulfoxylate (derivative of glyoxylic acid COHCOOH) $Na_2[O_2CCH(OH)SO_2]$ has been developed in Germany [Brueggemann Bruggolite] [80]. Details of synthesis have not been published.

2.3 Thiourea Oxides

One of the most important applications of thiourea (TU) is the preparation of thiourea dioxide, $(NH_2)_2CSO_2$ (TDO). Often manufacturers and chemists use other formula and names for TDO (NH_2NHCSO_2H, aminoiminomethanesulfinic acid, formamidinesulfinic acid), assuming the existence of SOH fragment in the compound. Indeed, as will be discussed later, the formation of such a compound (sulfinic acid) is possible in aqueous solution, but in solid state TDO exists exclusively in the form of $(NH_2)_2CSO_2$. By analogy, we will use the term "oxide" for TU mon- and trioxides as well (the other frequently used names are aminoiminomethanesulfenic acid, formamidinesulfenic acid or even sulfenyl acid [81] for thiourea monoxide and aminoiminomethanesulfonic acid and formamidinesulfonic acid for thiourea trioxide (TTO)).

TU oxides are synthesized by reaction of TU and hydrogen peroxide or peracetic acid at low temperatures (about 5°C) [1]. Trioxide can also be synthesized from dioxide. Use of 2 equivalent of hydrogen peroxide led to the dioxide (sulfinic acid), while use of 3 equivalent of that led to the trioxide (sulfonic acid):

$$(NH_2)_2CS + 2H_2O_2 \longrightarrow (NH_2)_2CSO_2 + 2H_2O, \qquad (2.14)$$

$$(NH_2)_2CS + 3H_2O_2 \longrightarrow (NH_2)_2CSO_3 + 3H_2O. \qquad (2.15)$$

In principle, for the oxidation of TU to its oxides other oxidants can also be used, for example, ferrate [81]. Oxides are the main products of TU oxidation by peroxides in neutral and weakly acidic solutions. In strongly acidic media (as well as in the presence of metal ions), reaction between TU and hydrogen peroxide mainly produces formamidine disulfide dication $(H_2N)(HN)CSSC(NH)(NH_2)^{2+}$ [1,82].

The first synthesis of TDO from TU and hydrogen peroxide in aqueous solution was reported in 1910 by Edward de Barry Barnett [37] (University College, London) in the paper "Action of Hydrogen Dioxide on Thiocarbamides". De Barry Barnett used neutral and alkaline solutions and synthesized a compound which, in his opinion, had a formula $NHC(NH_2)SO_2H$. Author did not use the name "thiourea dioxide". Instead of this he used the term "aminoiminomethanesulfinic acid". De Barry Barnett has also showed that the compound synthesized has reducing and acidic properties, fairly soluble in cold water and insoluble in organic solvents. He tried to get corresponding products from the reaction between allylthiourea and hydrogen peroxide. De Barry Barnett has determined a formula of synthesized product — $C_3H_5NC(NH_2)SO_2H$ or $C_3H_5NHC(NH)SO_2H$. This compound formed a viscous oil, which only crystallizes with utmost difficulty, and hence was not obtained in a state of purity. All attempts to obtain a pure product by oxidizing phenylthiourea, either in aqueous or in acetone solution, with hydrogen peroxide yielded a viscous oil which did not crystallize.

For many years TDO remained the only known oxide of TU, until Böeseken (Delft Institute of Technology, Netherlands) published papers on the chemistry of the TDO and trioxide [83,84]. He showed, that oxidation of TDO results in the formation of TTO, and named it as formamidinesulfonate. The most prolific group that studied thiourea oxides in 1950s–1970s was Walter's one from Hamburg University. A work from W. Walter's laboratory resulted in many publications on the chemistry of thioamides and thioureas, including papers on preparation and properties of thiourea mon- [85], di- [86] and trioxides [87,88]. Synthesis of TDO and TTO was also carried out by many other authors [39,89–97]. TDO can be received from TU and hydrogen peroxide, also directly at the place where TDO will be used, for example, as a bleaching agent [97]. The most convenient oxidant to prepare TTO from TDO is peracetic acid [96]. Despite the easiness of synthesis of TTOs, they are not produced in industrial scale and even by chemical companies producing reagents. The only industrial thiourea oxide is TDO. The main by-products of industrial production of TDO are TTO, urea, dicyandiamide, cyanamide, formamidine as well as unidentified compounds [98]. These substances show a wide range of different behavior in a biologic treatment process: while urea is known to be highly biodegradable, cyanamide and dicyandiamide are substances of low biodegradability and high nitrification inhibition. After oxidative chemical pretreatment, wastewater contains predominantly urea and sulfate together with minor amounts of dicyandiamide and ammonium [99]. It was demonstrated that this high strength urea wastewater can be treated biologically under aerobic conditions [99]. Due to the strong inhibition of nitrification by dicyandiamide, air stripping at elevated temperature and pH was found to be more efficient. This method offers the opportunity of transforming nitrogen into ammonium sulfate, which can be used as a raw material in the fertilizer industry.

The heat of reaction of TDO and hydrogen peroxide (≈ -536 kJ/mol) was calculated by Zhang and coworkers [100]. Some authors recommend using sodium molybdate as a catalyst of TU oxidation, which is frequently used in the hydrogen peroxide oxidation reactions [93,94].

An electrochemical procedure for the synthesis of TTOs has also been proposed [101]. These compounds were prepared by cathodic indirect oxidation of the corresponding TUs using a pertungstate/tungstate redox mediator driven by H_2O_2 electrogenerated from O_2 at the cathode and by the anodic indirect oxidation using a Cl_2/Cl^- redox mediator at the anode. A comprehensive list of synthesized TDOs and TTOs with the corresponding references can be found in the reference [1].

Thiourea monoxide is quite unstable and cannot be received in solid state [1]. However, formation of thiourea monoxide was observed during the course of oxidation of formamidinedisulfide (FDS) [102] and, possibly, TU [103] in solution. More stable monoxides are formed from TU derivatives containing bulky substituents at nitrogen atoms. Thus, N-phenyl- and ethylenethiourea (ETU) monoxides have been prepared by Ziegler and coworkers [104]. The formation of these compounds in solution was proved by the reaction with $FeCl_3$, typical of mono-S-oxygenated thioamides and thiocarbamates (oxidation of ETU will be discussed in detail later in this chapter). Recently valuable data have been received by Simoyi and coworkers. They studied oxidation of tetramethylthiourea by chlorite [105], bromine and bromate [106]. The oxidation pathway went through the formation of tetramethylthiourea sulfenic acid (tetramethylthiourea monoxide) as evidenced by electrospray ionization mass spectrum of the dynamic reaction solution. This S-oxide was then oxidized to produce tetramethylurea and sulfate as final products of reaction. There was no evidence for the formation of the sulfinic (dioxide) and sulfonic (trioxide) acids in the oxidation pathway.

In contrast to Simoyi's data [106], assuming the formation of tetramethylthiourea monoxide in the reaction between tetramethylthiourea and bromine, Li and coworkers have received in this reaction tetramethylformamidinedisulfide adding bromine dropwise to aqueous solution of tetramethylthiourea [107,108]. Authors used the redox couple tetramethylthiourea/tetramethylformamidinedisulfide in the dye-sensitized solar cells as an alternative to the conventional I_3^-/I^- redox couple. Advantages of this redox couple include its non-corrosive nature, low cost and easy handling. Besides, it operates well with carbon electrodes.

Attempts at synthesizing the trioxide as well as dioxide of trimethylthiourea made by Simoyi and coworkers failed [109] (authors do not mention what chemicals they used for oxidation of trimethylthiourea). There is also no direct evidence of formation of trimethylthiourea dioxide and trioxide during the course of reaction between trimethylthiourea and bromate in solution. Much earlier, however, Walter and Rohloff have received trialkylthiourea trioxides by the reaction between tetraalkylisothiouronium iodides and silver sulfite [88]. The analogous reaction of 1-chloro-N,N,N',N'-tetramethylformamidinium chloride with silver sulfite leads to the formation of tetramethylthiourea trioxide. Interestingly, Walter's group could not synthesize these trioxides by reactions of the corresponding TU with peroxides (in their opinion, due to instability of corresponding dioxides) [110,111]. Indeed, these two reactions (with participation of silver sulfite and peroxides) are quite different. In reaction with silver sulfite a new C–S bond is formed, as opposed to three atoms being transferred to the existing carbon-bound sulfur in case of the reaction with peroxides. Anyway, Walter's data show that tri- and tetraalkylthiourea trioxides can be synthesized and these compounds are relatively stable.

The mechanism of TU oxidation by different compounds leading to the formation of different products has been discussed in many papers including a recently published review [45,82]. As can be seen from the data mentioned above, the composition of products of reaction between TUs and H_2O_2 (or other peroxides) depend strongly on the structure of the corresponding TU and pH. The kinetics and mechanism of the oxidation of TU, N,N'-dimethyl (DMTU)- and N,N'-diethyl (DETU) TUs to their respective FDS cations by hydrogen peroxide in acidic media was studied by Hoffmann and Edwards [112]. This paper positively differs from many other papers on oxidation of thioureas since it presents detailed kinetic and mechanistic analysis as well as comparative kinetic data for different TUs. It has been determined that the oxidation of TU follows a two-term rate law (N — nucleophile):

$$\text{rate} = k_2[H_2O_2][N] + k_3[H_2O_2][N]H^+], \qquad (2.16)$$

where $k_3 > k_2$. Negative activation entropies are observed with $S^\#$ for TU>DMTU>DETU. The same order is observed for k_3 and for the apparent rate constants (k_{obs}). In their opinion, TUs act as nucleophiles by replacing the peroxide oxygen. Later, however, Saha and Greenslade [113] have shown that upon oxidation of TU by hydrogen peroxide the carbamidinothiyl radical $NH_2(=NH)CSC\bullet$ is formed, the most intense EPR signal being observed at pH 2.5–3.0.

TU's reaction with hydrogen peroxide in solution under conditions analogous to that using in the bleaching process with TDO has been investigated using ^{13}C-NMR spectroscopy by Arifoglu and coworkers [114]. This reaction is fast and exothermic, different intermediate products are formed during the course of reactions, depending on the pH of the medium and on the molar ratio of the reactants. The reaction of TU with hydrogen peroxide in hydrochloric acid at pH < 1 results in the formation of FDS dihydrochloride. In weakly acidic and neutral solutions (pH = 4.0–7.0) the reaction goes through a TDO intermediate. Because TDO hydrolyzes in solution and produces a strong reductant sulfoxylate (authors call it sulfinate), there is a rapid change in redox potential from a positive value to a high negative value.

Ziegler and coworkers [104,115] showed that flavin-containing monooxygenase of liver microsomes catalyzes the S-oxygenation of TU to the reactive electrophiles thiourea monoxide and TDO. Ziegler-Skylakakis and coworkers proposed that oxidized products of TU might be involved in genotoxicity [116]. Andrae and coworkers [117,118] hypothesized that the genotoxicity of TU may be the consequence of enzymatic oxidation of this compound.

Zhou and coworkers studied the reaction mechanism of the oxidation of TU by hydrogen peroxide in the gaseous state using density functional theory (BH&HLYP and B3LYP) and *ab initio* methods [119]. Their mechanism assumes the formation of two intermediates: aminoiminomethanesulfenic acid (intermediate 1) and its tautomer thiourea monoxide (intermediate 2). Both tautomers react with hydrogen peroxide yielding the final product aminoiminomethanesulfinic acid $(NH_2)(NH)CSO_2H$ (AIMSA), which is a tautomer of TDO (see Fig. 2.1).

Chatterjee and his coworkers have studied the oxidation of TU by H_2O_2 in the presence of the ruthenium complex, $[Ru^{III}(edta)(H_2O)]^-$

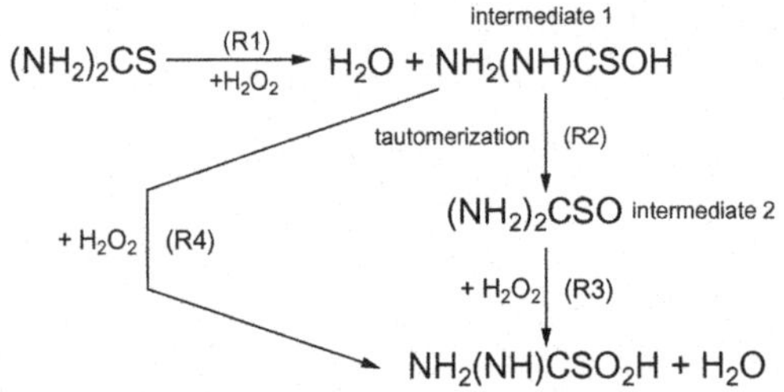

Fig. 2.1   Proposed mechanism of oxidation of thiourea by hydrogen peroxide

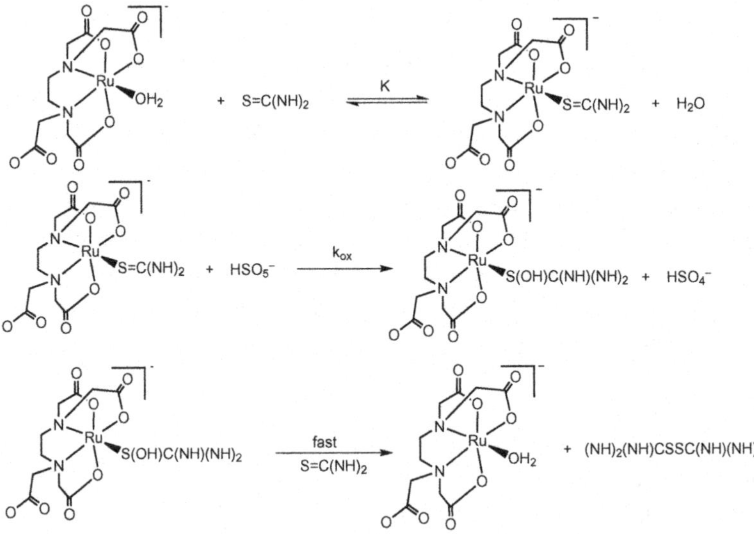

Fig. 2.2   Proposed mechanism for the Ru^{III}(edta)-catalyzed oxidation of TU by HSO_5^-

(edta^{4-} = ethylenediaminetetraacetate) at pH 4.9 [120]. HPLC product analysis revealed the formation of FDS as a major product at the end of the catalytic process; formation of other products, including TDO, TTO, and sulfate, was also observed after longer reaction times. The authors determined the rate constant for the conversion of FDS to TDO at 25°C, being 0.0058 $M^{-1}s^{-1}$. The same mechanism has been proposed recently by Sarkar and Chatterjee for the oxidation of TU by peroxomonosulfate ion (HSO_5^-) [121]. As proposed in Fig. 2.2, TU binds to ruthenium rapidly through the S atom (Eq. (2.17) in Fig. 2.2). Coordination of

TU through the S atom causes activation of the S=C bond, rendering it vulnerable toward direct attack by HSO_5^-. In the next rate-determining step oxidation of coordinated TU occurs, resulting in the formation of a S-hydroxylated $[Ru^{III}(edta)S(OH)C(NH)(NH_2)]^-$ intermediate species (Eq. (2.18) in Fig. 2.2). This species reacts rapidly with another molecule of TU to form formamidine disulfide $(NH:2)(NH)CSSC(NH)(NH_2)$ in a kinetically indistinguishable step (Eq. (2.19) in Fig. 2.2). Formamidine disulfide is not stable under condition used (pH = 6.2). In this study, authors have also noticed complete depletion of FDS with concomitant formation of TDO. The findings of this work taken together with those reported earlier [121], strongly suggest that the oxidation of S-bonded TU in $[Ru^{III}(edta)(TU)]^-$ takes place via oxygen-transfer from the oxidant to the S atom of the coordinated TU in $[Ru^{III}(edta)(TU)]^-$ via heterolytic cleavage of the O–O bond, which is kinetically highly favored for HSO_5^-. Thus, Chatterjee's mechanism assumes formation of FDS not in the course of radical reactions.

Gao and coworkers, based on HPLC data on the oxidation of FDS and TU by hydrogen peroxide [102,103] proposed that in strongly acidic solutions (pH = 1.50) the first intermediate of reaction between TU and H_2O_2 is aminoiminomethanesulfenic acid ($k = 0.115$ $M^{-1}s^{-1}$), but not FDS, and that the latter compound is formed from the following reaction (Eq. (2.20)) with a normal second-order reaction with a rate coefficient of $k = 4.5$ $M^{-1}s^{-1}$:

$$(NH_2)(NH)CSOH + (NH_2)_2CS \longrightarrow NH_2NHCSSCNHNH_2 + H_2O.$$

$$(2.20)$$

This reaction is considered by authors to be an irreversible process. Data of Chatterjee and his coworkers, who observed formation of TU oxides after FDS, however showed that it can be reversible and, at least in acidic solutions and/or in the presence of complexes or salts of redox active metals (their presence favors radical process), FDS can be treated as a primary source of thiourea monoxide and then further, with excess peroxide, as a source of di- and trioxide.

All the mentioned data of interconversion of FDS and thiourea monoxide are related to acidic solutions. Since FDS is unstable even in weakly acidic media [102], there are no data on formation of FDS as an intermediate of reaction between TU and peroxides in neutral and alkaline solutions. Most probably, in these solutions in the absence of complexes or salts of redox active metals, TU acts as a nucleophilic agent by replacing one of the oxygen atom of peroxides in a simple, formal, oxygen-transfer process.

Data mentioned above on the formation of FDS as a final product of radical oxidation of TU have been confirmed by Wang and coworkers [122]. They studied •OH-radical induced oxidation of TU and tetramethylthiourea. The only primary product of reaction is FDS (tetramethylformamidinedisulfide).

TU and allylthiourea are known to undergo of dye-sensitized oxidation to give corresponding dioxides (sulfinic acids) [123]:

$$R-NH-C(S)-NH_2 \xrightarrow{h\nu/O_2,\text{dye}} R-NH-C(NH)-SO_2H, \qquad (2.21)$$

where R stands for H or $CH_2CH=CH_2$ group. This process has been confirmed to be a singlet oxygen reaction at the TU group by a kinetic study [124]. Later, TDO was identified as a product of reaction between TU and singlet oxygen in ethanol containing rose bengal [125]. However, there was evidence of further reaction by the identification of sulfur, sulfur dioxide and dicyandiamide. Substituted TUs (N-methyl-, N-ethyl-, N-phenyl-, N,N'-dimethyl-, N,N'-diethyl, N,N'-diphenyl-, ethylene-) do not produce corresponding dioxides (sulfinic acids) [125]. In contrast to the thioamides, none of the TUs studied underwent either photolysis or photooxidation in the absence of sensitizers.

Superoxide ion, in aprotic media, converts TU and monosubstituted thioureas into cyanamides [126]. It is proposed that superoxide attacks TUs to form dioxides, which after elimination of sulfoxylate ion SO_2H^- produce cyanamides. Reactions of superoxide (KO_2) with diarylthioureas produce triarylguanidines and sulfate ion [127]. Interestingly, again, as in the case of TU and monosubstituted TUs, authors assume the formation of dioxide as the intermediate, although the mechanism of triarylguanidine formation is not clear.

Reaction of TU with ferrate has been studied by Sharma [81]. A mechanism was proposed based on the stepwise oxidation of sulfur from the oxidation state of -2 to $+6$ by Fe(VI). It was postulated that Fe(VI) is reduced to Fe(V) by TU with the formation of a radical $NH_2NHCS\bullet$. The radical further reacts with Fe(VI) to form Fe(V) and thiourea monoxide (author calls it sulfenyl acid). It rapidly reacts with Fe(V) to give TDO, which, in turn, reacting with Fe(V), gives TTO. Since the reactivity of Fe(V) is 3–5 orders of magnitude higher than that of Fe(VI), thiourea monoxide and TTO will preferentially react with Fe(V) rather than Fe(VI). The subsequent oxidation of TTO accompanied by C–S bond cleavage gives urea and sulfate. Indeed, the experiments with the boiler chemical cleaning

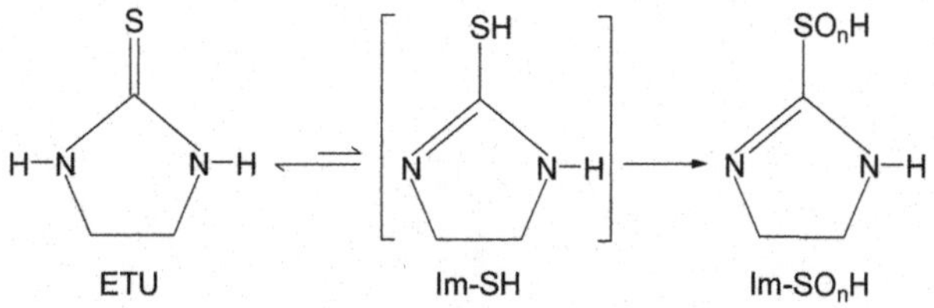

Fig. 2.3 Oxidation of ETU to the corresponding sulfenic (n = 1), sulfinic (n = 2) and sulfonic (n = 3) acids

wastes (BCCW) showed that the stoichiometric excess of Fe(VI) completely removed TU from BCCW samples.

Among the substituted TUs one of the most important one is ETU (imidazolidine-2-thione, Im-SH). ETU is a carcinogenic degradation product of major ethylene bis(dithiocarbamate) fungicides [128]. The processes of oxidation of ETU have been studied in detail [128]. Thus, the reaction of ETU with H_2O_2 was examined in aqueous medium at pH = 5, 7 and 9 by ^{1}H-NMR spectroscopy giving five principal products: sequential formation of sulfenic (Im-SOH), sulfinic (Im-SO$_2$H), and sulfonic (Im-SO$_3$H) acids (Fig. 2.3) as well as imidazoline and ethyleneurea. Maximum yields with 2 equiv. of H_2O_2 at optimal pH were 10%, 71%, 5%, 53% and 100%, respectively. Oxidation proceeds mainly through the sulfinic acid to imidazoline in acidic medium and the S-oxide to ethyleneurea in basic medium [128].

Thus, the above-mentioned data discussed show that the processes of TUs oxidation by different oxidants are very different and composition of their products strongly depends on the structure of TUs. In many of these processes, TDO is a final or intermediate product.

Chapter 3

Structure

3.1 Dithionites

Despite the apparent simplicity, the structure of dithionites, hydrox-ymethanesulfinates and thiourea oxides in solid states and solutions aroused a great interest and was discussed in many papers. It has been studied in detail by UV–Vis, IR, Raman and EPR spectroscopy, X-ray diffraction analysis and density functional (DFT) calculations. The earliest studies were dedicated to structure of sodium dithionite. According to X-ray diffraction data [129], the $S_2O_4^{2-}$ ion belongs to the C_{2v} point group and contains an abnormally long S–S bond (2.389 Å). In the cage-like tin(II)-dithionite $Sn_2(S_2O_4)_2$ complex ion Sn^{2+} is coordinated to four oxygens from two $S_2O_4^{2-}$ ions. The dithionite ion has a slightly distorted C_{2v} configuration with a S–S distance of 2.350 Å [62]. In the other stable compound, $ZnS_2O_4 \cdot$ pyridine, S–S distance is almost the same as in the sodium salt (2.386 Å) [130]. X-ray diffraction data on tetraethylammonium dithionite are unavailable [63]. Solid-state EPR studies indicate that the isolated solid $[Et_4N]_2[S_2O_4]$ contains ca. 1% of $[Et_4N][SO_2]$. The solubility of $[Et_4N]_2[S_2O_4]$ in water and organic solvents (DMF, DMSO and acetonitrile) gave a possibility to compare structure of dithionite in aqueous and non-aqueous solutions. Indeed, at relatively small concentrations (1.02–1.61 mM) the 313 nm band, which is a characteristic of the dithionite ion in H_2O (molar absorptivity is 8043 $M^{-1}cm^{-1}$[58]), does not appear in the spectrum of $[Et_4N]_2[S_2O_4]$ in organic solvent. The explanation of this behavior is that dithionite anion is much more dissociated to sulfur dioxide anion radical in organic solvents than in aqueous solution (see Eq. (3.1)):

$$S_2O_4^{2-} \rightleftharpoons 2 \cdot SO_2^-. \tag{3.1}$$

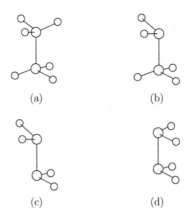

(a) (b)

(c) (d)

Fig. 3.1 Observed structure of dithionate (a), $S_2O_5^{2-}$ (b), predicted structure of dithionite (c) and dithionite (d)

Thus, organic solvent does not influence structure of dithionite ion itself, but increases concentration of sulfur dioxide anion radical.

The other possible factor that can influence the structure of dithionites is a type of cation. The authors of the papers [131,132] have shown by Raman spectroscopy that after dissolving of $Na_2S_2O_4$ in water structure of dithionite ion changes, it becomes centrosymmetrical and corresponds to the C_{2h} point group (Fig. 3.1). It also show [133] that the abnormal eclipsed conformation (C_{2v}) of the solid sodium dithionite is a characteristic of salts with small cations, for example, with sodium. In the salts with large cations, for example, with tetraethylammonium, dithionite is centrosymmetrical both in the solid state and in solution. Thus, the eclipsed conformation of $Na_2S_2O_4$ in the solid state is due to the influence of the cation and associated packing effects [133] rather than to the properties of dithionite ion itself. Structure of sodium dithionite in the solid state is quite different from the structures of other sodium salts of sulfur-containing anions with S–S bonds, e.g. $S_2O_5^{2-}$ and $S_2O_6^{2-}$ [133], which adopt staggered configurations both in solid state and in solution, with $S_2O_6^{2-}$ belonging to centrosymmetric point group D_{3d} (Fig. 3.1). Experimental data were confirmed by the results of LCAO-Xα DFT calculations [134]. It was shown that the curve for the total energy vs. S–S bond length for dithionite ion, unlike for $S_2O_5^{2-}$ and $S_2O_6^{2-}$, has a very broad minimum. This indicates that this bond length in dithionite ion depends strongly on external factors, for example, packing effects in the crystal.

Raman spectra of sodium dithionite dihydrate reveals that the dithionite ion is in a different conformation than in the anhydrous material [135]. S–S bond length in a dihydrate (2.298 Å) is substantially shorter than in the anhydrous structure and an O–S–S–O torsional angle is 56°, approximately gauche. In the anhydrous sodium dithionite O–S–S–O torsional angle is 16°.

The existence of at least two forms in the solid state are also characteristics of lithium dithionite [136]. It was shown on the base of Raman spectroscopy studies that lithium dithionite prepared under rigorous non-aqueous conditions has properties significantly different from those of $Li_2S_2O_4$ prepared in aqueous environment. Thus, according to data of differential scanning calorimetry, $Li_2S_2O_4$ (aqueous) exhibit a large characteristic exotherm at 468 K, while the large exotherm for $Li_2S_2O_4$ (non-aqueous) occurs at 445 K. Interestingly, that $Li_2S_2O_4$ (non-aqueous) obtained in the scaled-up reaction was extremely susceptible to oxidation, burning spontaneously upon exposure to air. The type of lithium dithionite (non-aqueous) is not observed after the material is reprecipitated from aqueous solution. Structural data on different forms of lithium dithionite are unavailable.

The structure of dithionite is so sensitive that it responds not only to transfer from solid state to solution, but also to other external influences, for example, coordination. The first structurally characterized complex of a sulfur-bound dithionite ion (C_{2h} symmetry) is $[Cp^*Mo(CO)_3]_2\mu\text{-}S_2O_4$), where $Cp^* = \eta^5\text{-}C_5Me_5$ (Fig. 3.2) [69]. It exhibits a S–S bond distance of 2.266 Å, which is 0.12 Å shorter than that in $Na_2S_2O_4$. But this distance is significantly longer than that found for diphenyl disulfone, $Ph_2S_2O_4$ (2.193 Å) [137], the other structure for which the S_2O_4 moiety is found in approximate C_{2h} symmetry. It should be noted that similar dithionite complex $[\eta^5\text{-}C_5H_5Fe(CO)_2SO_2]_2$ has been synthesized earlier by Wojcicki and coworkers [138–140] but in their papers X-ray data are absent.

Japanese researchers obtained binuclear rhodium complex in a crystalline form $[(RhCp^*)_2(\mu\text{-}CH_2)_2(\mu\text{-}O_2SSO_2)]$ (1), containing the photoactive ligand dithionite ($\mu\text{-}O_2SSO_2$) and two pentamethyl-cyclopentadienyl ligands [26,27] (the prefix μ in the above formulas denotes a bridging ligand and the prefix η indicates the hapticity of the ligand). The $\mu\text{-}O_2SSO_2$ ligand in 1 is coordinated parallel to the Rh–Rh bond and has a weak S–S bond 2.330 Å. This is the first example of a side-on type coordination mode for dithionite ion. Under the action of light, the complex $[(RhCp^*)_2(\mu\text{-}CH_2)_2(\mu\text{-}O_2SOSO)]$ (2) is obtained in a 100% yield in a reaction accompanied by

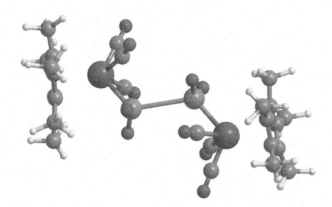

Fig. 3.2   ORTEP projection of $[Cp^*Mo(CO)_3]_2(\mu\text{-}S_2O_4)$

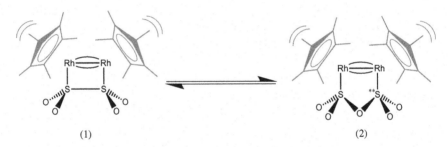

Fig. 3.3   Reversible photochromic transformation of $[(RhCp^*)_2(\mu\text{-}CH_2)_2(\mu\text{-}O_2SSO_2)]$ (*: asymmetric sulfur atom)

a color change, where instead of an S–S bond from dithionite an S–O–S moiety is formed (Fig. 3.3).

It must be emphasized that compounds such as the sulfur oxide O_2SOSO were not known until the reports of Nakai and coworkers [26,27]. When stored in the dark at room temperature for 3 weeks, compound (2) completely reverts to (1). This reversible photochromic reaction in the crystalline state can also be repeated. Since all other photochromic reactions known to date occur in crystals with yields not exceeding 15%, binuclear rhodium complexes with the dithionite ligands have good prospects for use as functional molecular crystals [26–29].

Structure of the Yb^{III} dithionite/sulfinate complex $[\{(\eta^5\text{-}C_5Me_5)_2 Yb(\mu_3,1\kappa^2O^{1,3},\ 2\kappa^3O^{2,2',4}\text{-}S_2O_4)\}_2\{(\eta^5\text{-}C_5Me_5)Yb(\mu,1\kappa O,2\kappa O'\text{-}C_5Me_5 SO_2)\}_2]$ (3) and Yb^{III} dithionite complex $[\{(\eta^5\text{-}C_5Me_5)_2Yb\}_2\ (\mu,1\kappa^2O^{1,3},$

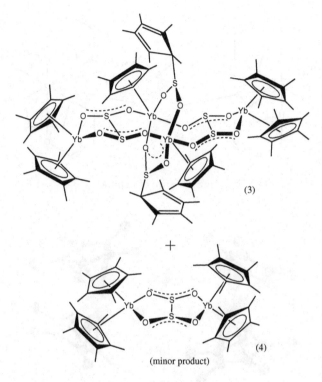

(3)

+

(4)

(minor product)

Fig. 3.4   Structure of complexes 3 and 4

$2\kappa^2O^{2,4}$–$S_2O_4)]$ (4) (see Fig. 3.4) has been studied by Klementyeva and coworkers [24].

Dithionite complex of tetravalent uranium [{(np,MeArO)$_3$tacnU^{IV}}$_2$ $(\mu$-κ^2: κ^2– $S_2O_4)]$ (tacn=triazacyclononane) (5) differs significantly from most literature-known dithionite complexes (Fig. 3.5) [23]. Due to the oxophilicity of the uranium center, uranium forms complex, in which the dithionite ligand bridges through the sulfur instead of the oxygen atoms. The complex exhibits a S–S bond distance of 2.346 Å.

The same group studied the structure of samarium dithionite complex [{(η^5-C$_5$Me$_5$)$_2$Sm}$_2$(S$_2$O$_4$)] (6) (Fig. 3.6) [25]. As in uranium complex, in this complex the S$_2$O$_4^{2-}$ anion binds side-on to each metal atom through two oxygen atoms. The S–S bond length (2.420 Å) in this complex is longer than the complexes mentioned above. In the dithionite–sulfinate complex [{(η^5-C$_5$Me$_5$)$_2$Sm(S$_2$O$_4$)}$_2${(η^5-C$_5$Me$_5$)Sm(C$_5$Me$_5$SO$_2$)}$_2$] this bond length is slightly shorter (2.369 Å).

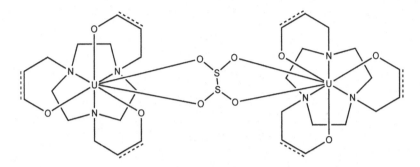

Fig. 3.5   Structure of complex (5)

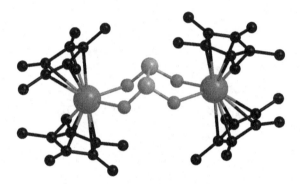

Fig. 3.6   Structure of complex (6). Hydrogen atoms are omitted for clarity

In the end of review of dithionite structure, note that this ion has one more interesting feature. By means of quantum chemical calculations Zhang and coworkers showed that $S_2O_4^{2-}$ has aromaticity resulting from through-space conjugation, analogous to the aromaticity in the transition states of some pericyclic reactions, such as Cope rearrangements and Diels–Alder reactions [141].

3.2   Hydroxymethanesulfinates

The crystal lattice parameters of sodium hydroxymethanesulfinate dihydrate and the interatomic distances in the $HOCH_2SO_2^-$ anion were determined by Truter [142,143]. Before appearance of her works, the constitution of this anion was uncertain. The analysis showed that the sulfur is present in the oxidation state IV and pyramidally bonded to two oxygen atoms

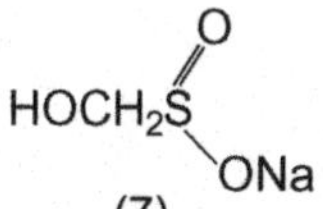

$$\text{HOCH}_2\overset{\displaystyle O}{\underset{\displaystyle \diagdown ONa}{S}}$$

(7)

HOCH₂O—S—ONa

(8)

Fig. 3.7   Possible structures of hydroxymethanesulfinate in aqueous solution

(S–O bond length is 1.50 Å) and one carbon atom with a long S–C bond, 1.84 Å, and therefore the anion should be written $HOCH_2SO_2^-$.

However, many of the chemical data available on the reactions of hydroxymethanesulfinate conflict with this tetravalent sulfur structure (see (7) in Fig. 3.7) [144]. The data which conflict with this structure most seriously are those from redox titrations. Possible explanation of this conflict is the existence of the other form of hydroxymethanesulfinate (with O–S–O fragment, (see (8) in Fig. 3.7) in aqueous solution [144]. This can account for the +II oxidation state of sulfur inferred from data from redox titrations.

However, the results obtained later by IR and Raman spectroscopy strongly suggest that "sodium formaldehyde sulfoxylate" has the sodium hydroxymethanesulfinate structure $HO–CH_2–SO_2Na$ having the C–S linkage both in the solid states and aqueous solutions [145]. Thus, conflict with data of redox titrations remain unresolved. In principle, there are two possible explanations of that. The first one is the formation of sulfoxylic acid $S(OH)_2$ from hydroxymethanesulfinate (see later), the second one is a rapidly established equilibrium between two forms of this compound (7 and 8) (concentration of form with O–S–O fragment is very low and therefore it does not manifest in IR and Raman spectra).

In contrast with the structure of dihydrated sodium hydroxymethanesulfinate where the organic residue, being involved in an extensive network of hydrogen-bond contacts, is fully stretched and bridges two Na ions through a single O-atom, zinc hydroxymethanesulfinate possesses, in the solid state, a polymeric framework, with the Zn atoms octahedrally coordinated by six O atoms of four different hydroxymethanesulfinate groups (see Figs. 3.8 and 3.9). The latter are found to coordinate through all their O atoms, including the hydroxy ones, and possess a chelating nature [146].

3.3   Thiourea Oxides

First X-ray study of thiourea dioxide (TDO) was performed by Sullivan and Hargreaves [147]. Later the data were refined [148–150]. In the solid state

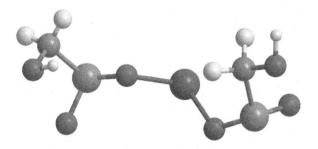

Fig. 3.8   Structure of zinc hydroxymethanesulfinate

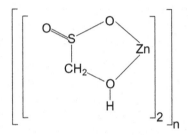

Fig. 3.9   Polymer of zinc hydroxymethanesulfinate

TDO exists in the form of $(NH_2)_2CSO_2$, with a pyramidal CSO_2; the S–O and C–N bond lengths are 1.496 Å and 1.296 Å, respectively, and the C–S bond (1.867 Å, see Table 3.1) is much longer than in thiourea (1.716 Å). A longer C–S bond was reported for the N,N′-dimethylthiourea dioxide (1.880 Å) [151]. The C–S bond lengthening is likely due to the antibonding interaction between the filled p-orbitals in the CN_2 central unit and the sulfur lone pair [151].

Using X-ray data, Chen and Wang [148] have concluded that the C–S bond in TDO is essentially a single bond. Song and coworkers [150] suggested that TDO is a combination of two forms (Fig. 3.10).

A highly polar single C–S bond between two oppositely charged fragments has been established [152]. The large dipole moment results from a positive $C(NH_2)_2$ moiety and a negative SO_2 fragment, a firm illustration of the zwitterionic character [152]. However, in 2003, Denk and his coworkers

Table 3.1   Selected bond distances (Å) of some thiourea oxides

Compound	C–S	S–O(1)	S–O(2)	S–O(3)	N(1)–C(1)	N(2)–C(1)	Reference
$(NH_2)_2CSO^a$	1.704	1.553			1.372	1.345	[156]
$(NH_2)_2CSO_2$	1.867	1.496	1.496		1.296		[150]
$(N^{(2)}H)CH_3N^{(1)}$ $H_2CSO_2 \cdot H_2O^b$	1.860	1.499	1.471		1.293	1.307	[157]
$(NHCH_3)_2C^{(1)}$ SO_2	1.880	1.479	1.476		1.303	1.304	[151]
$(NH_2)_2CSO_3$	1.815	1.446	1.431	1.439	1.298	1.297	[156]
$(NHCH_3)_2C^{(1)}$ $SO_3 \cdot H_2O^c$	1.820	1.433	1.439	1.441	1.290	1.308	[157]

[a] Calculated using B3LYP method with 6-311+ G(d,p) basis set.
[b] Longer bond with sulfur atom forms oxygen atom, which forms a bond with water.
[c] Atom $O^{(2)}$ forms a bond with water.

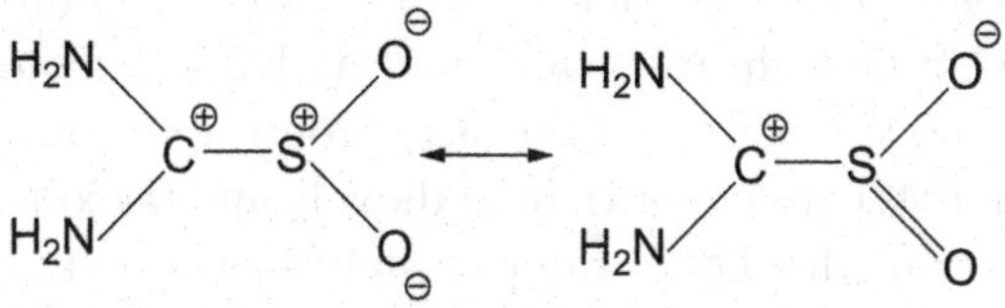

Fig. 3.10   Redox isomers of TDO

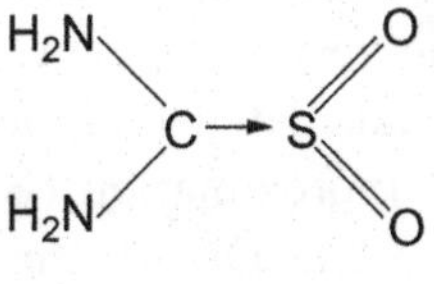

Fig. 3.11   Carbenoid structure of TDO

[153] presented an experimental evidence and quantum chemical calculations indicating that thiourea dioxides (TDO) are Lewis acid–base adducts of diaminocarbenes and described TDO structure as shown in Fig. 3.11.

The zwitterionic and carbenoid structures of TDO were compared by Kis and coworkers [154]. Calculation of NBO (natural bond orbital) partial atomic charges on TDO showed that there is only a small charge separation between the two moieties of TDO ($C(NH_2)_2$ and SO_2); i.e. TDO does not appear to be well described by a zwitterionic structure. A comparison of C–S bond length in TDO with C–S bond lengths in other sulfur-containing

compounds with well-defined structure (methanethiol and thioformalde-
hyde) showed that the C–S bond in TDO is significantly longer/weaker
than a normal C–S single bond.

Structure of thiourea trioxide were studied by Makarov and coworkers [155]. The molecular geometry about the central C atom was found
to be strictly planar. The S–O bonds vary from 1.431 Å to 1.446 Å, the
C–N bonds are virtually equivalent (Table 3.1) and are much shorter than
the typical C–N bond (1.470 Å). The C–S bond (1.815 Å) is less than in
TDO [155]. Note that double methylation of thiourea increases the C–S
bond both in trioxide and dioxide, but in N-methylthiourea dioxide this
bond is shorter than in dioxide of unsubstituted thiourea (Table 3.1).

Due to instability of thiourea monoxide (TMO) experimental structural
data for this compound are unavailable. Some information on its structure
can be found only in the paper of Peng and coworkers who performed
a theoretical study on interaction between TMO and water [156]. They
estimated the C–S bond in this compound as 1.704 Å which is much less
than in dioxide and trioxide (see Table 3.1). Contrary to this, S–O and C–N
bond lengths in TMO are larger than in dioxide and trioxide. The C–S and
S–O bond lengths of the TMO-water clusters increase gradually with the
addition of water molecules, showing that the strength of the C–S and S–O
bonds has weakened, and indicating that monomer TMO becomes more
unstable with increasing water molecules.

Considering the structural features of thiourea oxides, it is absolutely
mandatory to emphasize the very important stabilizing role of hydrogen
bonds. Fang and coworkers calculated bond lengths for TDO using DFT
within generalized gradient approximation (GGA) and the local density
approximation (LDA) [158]. The O$\cdots$H lengths vary from 1.747 Å to
2.061 Å, indicating that there exist strong H–bonds in the TDO crystal. These values are close to experimental bond lengths (1.821 Å and
2.060 Å) [148]. Both GGA and LDA results show that each TDO molecule
is involved in eight hydrogen bonds between four neighboring molecules.

The relative stability of different clusters of TDO in water was examined using gas phase quantum chemical calculations [159]. The calculation
results showed that a strong interaction exists between TDO and water
molecules, as indicated by the binding energies of the TDO clusters pro-
gressively increased by adding water molecules.

The crystal structure and hydrogen-bonded networks at ambient pressure have been studied recently by Shao and coworkers [160]. In this work,

TDO, AIMSA, and 12 TDO or AIMSA clusters are simulated to get insight into the structure and properties of TDO solubilized in water. Authors have considered various hydrogen-bond interactions with participation of the lone pairs on water, oxygen and nitrogen atoms of TDO and AIMSA for accepting hydrogen bonds. Their calculations showed that TDO more strongly interacts with itself than with water. In the authors' opinion, this phenomenon explains relatively low solubility of TDO in water. It was also found that when TDO solubilizes in water, formation of the cyclic structures constructed by water and TDO is more possible than the formation of linear structures. Interestingly, that though authors of this paper emphasize significant changes of TDO structure after solvation by water, they do not connect them with tautomerization of TDO to AIMSA but just with formation of TDO-water oligomers with cyclic structure.

The possibility of such tautomerization was studied by Makarov and Kudrik [161]. Using quantum chemical calculations they showed that in aqueous solutions aminoiminomethanesulfinic acid (AIMSA) NH_2NHCSO_2H is a more stable form, than $(NH_2)_2CSO_2$, i.e. tautomerization of TDO into AIMSA after dissolving the solid material is possible. They suggested an intermolecular mechanism of tautomerization. Later, intermolecular and intramolecular tautomerization mechanisms were both found to be feasible for interconversion of TDO to AIMSA [154].

The crystal structure and hydrogen-bonded networks in TDO at high pressure have been studied by two independent research groups [162,163]. Davidson and coworkers have reported a phase transition between a powder sample of orthorhombic phase I of TDO to a new monoclinic phase II at a pressure of 0.54 GPa [162]. This transition has also been observed in a single crystal sample at a pressure of 0.45 GPa. They have also reported an unusual isostructural transformation in TDO at 6.8 GPa that involves the formation of a new hydrogen bond. Later Wang and coworkers observed marked changes in the Raman spectra of TDO at 3.7 GPa, which strongly indicated a structural phase transition associated with the distortions of hydrogen bonding [163]. There were no further changes up to the maximum pressure of 10.3 GPa and the observed transition was completely reversible when the system was brought back to ambient pressure. This transition was further confirmed by the changes of angle-dispersive X-ray diffraction (ADXRD) spectra. The results from the first-principles calculations suggested that this phase transition was mainly related to the changes of hydrogen-bonded networks in TDO.

Thus, data discussed in this chapter indicate that structures of dithion-ites, hydroxymethanesulfinates and thiourea oxides are very unusual and sensitive to external influences. Despite the sometimes contradictive con-clusions, especially concerning structure of TDO, structural data are very helpful in analysis of properties of these sulfur-containing compounds.

Chapter 4

General Properties and Analysis

Commercially available sodium dithionite $Na_2S_2O_4$ is a white powder with a purity of approximately 88%. Main impurities are sodium sulfite, metabisulfite and thiosulfate [164]. Double recrystallization of commercial dithionite from 0.1 M NaOH–methanol under anaerobic conditions gives a product with a purity of 99 ± 1% [58]. Dithionite also forms dihydrate $Na_2S_2O_4 \cdot 2H_2O$. Sodium dithionite dihydrate is very sensitive toward atmospheric oxygen in the finely crystalline state and it is oxidized under heat development, therefore, all further information refers to anhydrous sodium dithionite [164]. Sodium dithionite is highly soluble in water (182 g/L at 20°C) [164], but almost insoluble in organic solvents just as all the main group metal dithionites [60]. The other dithionite salt, $[(C_2H_5)_4N]_2S_2O_4$, is soluble in dimethylformamide, dimethylsulfoxide and acetonitrile [63]. The purity of lithium dithionite $Li_2S_2O_4$ is of 90–92% [59,165], potassium dithionite contains less than 70% $K_2S_2O_4$ [60]. The density and viscosity of aqueous solutions of sodium dithionite, sodium dithionite + sucrose and sodium dithionite + sodium hydroxide + sucrose over the temperature range (25–40°C) were measured by Vázquez and coworkers [166]. Data on toxicity and effects on human health are summarized in Ref. [164]. Particularly, the acute oral LD_{50} of sodium dithionite in rats is about 2500 mg/kg bw.

Thermal decomposition of sodium dithionite were studied by Erdey and coworkers [167] as well as by Flaherty and Bather [168]. It was found that at 170°C, in exclusion of air, vigorous decomposition occurs, yielding sodium thiosulfate, sodium sulfite and sulfur dioxide:

$$2Na_2S_2O_4 \longrightarrow Na_2S_2O_3 + Na_2SO_3 + SO_2. \qquad (4.1)$$

The most well-known analytical method of determination of dithionite is iodometric titration. In 1923, Merriman described a simple titration procedure to quantify dithionite which required no inert atmosphere [169]. The first part of this method involves the reaction of a sample of dithionite with formaldehyde, producing sodium hydroxymethanesulfinate (rongalite, HMS) $HOCH_2SO_2Na$ and hydroxymethanesulfonate $HOCH_2SO_3Na$:

$$Na_2S_2O_4 + 2CH_2O + H_2O \longrightarrow HOCH_2SO_2Na + HOCH_2SO_3Na. \quad (4.2)$$

The first product can be titrated with iodine in weakly acidic solutions, but the second product cannot be:

$$HOCH_2SO_2Na + 2I_2 + 2H_2O \longrightarrow NaHSO_4 + 4HI + CH_2O. \quad (4.3)$$

Obviously, Merriman's method can give satisfactory results only if sulfite is the sole reducible species in the dithionite sample. In the later studies, authors tried to take into account other sulfur-containing species (thiosulfate, sulfide) [170–177]. Wollak detailed a series of three iodometric titrations which eliminated the interferences from additional reducible species and measured the amounts of dithionite, thiosulfate and sulfite (i.e. bisulfite) present in the samples [170]. However, other authors have mentioned that Wollak's method is unsatisfactory if thiosulfate and sulfite are present in high concentrations [171,173]. Kilroy has published a series of papers in order to revise Wollak's method for dithionite analysis [174–177]. The Wollak's procedure for the determination of the sum of dithionite and thiosulfate has been investigated. Analytical data and a kinetic study confirm that thiosulfate reacts with formaldehyde under acidic conditions. Both pH and CH_2O concentration affect the results and satisfactory analysis can be performed only if the pH of the initial sample — formaldehyde mixture remains high enough [174,175]. In his final paper, Kilroy has reported a method of analyzing mixtures of soluble sulfides, thiosulfate, dithionite and sulfite. The selected method permits analysis of heterogeneous samples or samples not conveniently separated. The scheme combines several methods, including the iodometric procedure for analysis of inaccessible mixtures containing dithionite, modification of the dithionite determination with methylene blue and the analysis of soluble sulfides by the iodate method [177].

Besides iodometric methods, other analytical procedure for determination of dithionite has also been reported. A rapid titration procedure was recommended for the estimation of $Na_2S_2O_4$ in commercial sodium dithionite and in aqueous solutions of this compound, using potassium hexacyanoferrate(III) as a reagent and methylene blue as an indicator [178]. With the procedure described, there is no interference of decomposition products of dithionite (sulfate, sulfide, sulfite, trithionate and thiosulfate).

Alternate methods for dithionite composition analysis are also used for dithionite determination. Reported methods for dithionite quantification include mainly electrochemical ones [179] (polarography [180–185], chronoamperometry [186,187], cyclic voltammetry [188], capillary zone electrophoresis [189] and isotachophoresis [190]) as well as chromatography [191–196], Raman spectrometry [193,197], and spectrophotometry [198]. Polarographic methods are well known for oxo-sulfur species and were used not only for quantification of dithionite, but also for determination of the products of its decomposition. The main drawback of early polarographic methods was an interference of sulfide. In such cases, the interference must be eliminated by the addition of zinc sulfate to the supporting electrolyte to precipitate the sulfide ion [180]. De Carvalho and Schwedt have employed differential pulse polarography (DPP) to analyze dithionite solutions [183,184]. Using different supporting electrolyte solutions the determination of dithionite, thiosulfate, sulfide and elemental sulfur without interferences was shown to be possible. Recently, DPP was successfully applied to determine the dithionite content of sugar and loaf sugar samples [185].

Chronoamperometric method was used to measure continuously and simultaneously the concentrations of sodium dithionite and dyes sulfite for application in textile dyeing processes [186,187]. It was shown in the earlier study with rotating disc electrode that the sodium dithionite concentration can be monitored continuously using chronoamperometry but only in the absence of dyestuff due to its adsorption [186]. Later the same group has found that the oxidation reaction of indigo, sodium dithionite and sulfite at a platinum electrode in a wall-jet configuration can be used for their simultaneous detection because of the difference of the half-wave potentials of their voltammetric waves [187]. Implementation of a wall-jet instead of a rotating disc electrode is much easier and cost effective. Sodium dithionite is oxidized at a potential of 0.3 V (here and further in this section vs. Ag/AgCl). Indigo is oxidized at -0.55 V to a virtually water insoluble product, which precipitates at the electrode surface. Indigo behaves quasi-reversibly, and is reduced at a potential of -0.9 V. In order to clean the electrode surface this reduction is used as a step in the multistep sequence [187]. Two years later, it was seen that the amperometric detection of micro-molar of dithionite is possible at electrodeposited nickel oxide film on carbon ceramic electrodes [188].

De Carvalho and Schwedt have developed a capillary zone electrophoretic method for the separation and following determination of the dithionite (as $HOCH_2SO_2^-$ and $HOCH_2SO_3^-$ anions in the presence of formaldehyde), sulfite (as $HOCH_2SO_3^-$ anion), sulfate and thiosulfate. The method allowed the determination of dithionite, sulfite, sulfate and thiosulfate in commercial formulations of bleaching agents [189].

Isotachophoretic determination of dithionite and metabisulfite in technical samples has been suggested by Nováková and coworkers. Similar to the method of de Carvalho and Schwedt [189] dithionite and metabisulfite present in the samples were transformed by the reaction with formaldehyde to stable compounds, hydroxymethanesulfinate and hydroxymethanesulfonate that were determined isotachophoretically without any pretreatment except for sample filtering and degassing [190].

A series of papers has been dedicated to different chromatographic methods of dithionite analysis [191–196]. Separation techniques include both ion exchange and ion-pair chromatography. The most widely used detection system is the conductometric [191–193] or UV absorbance [192–194] detectors. James and coworkers have analyzed five sodium dithionite samples using iodometric titration method [193] and an ion chromatography (IC) approach [192,193]. Samples for IC analysis were prepared by adding a sample of dithionite to a basic formaldehyde solution. James *et al.* have shown that the IC method provides a simple one-step protocol to rapidly and accurately determine the concentration of dithionite. The results were in excellent agreement with those obtained by using a multistep iodometric titration. Compared to the titration method, the IC approach requires less solution preparation and sample analysis and the results can be rapidly obtained. Steudel and Munchow have demonstrated that dithionite can be determined chromatographically after derivatization to the more stable hydroxymethanesulfinate, which has absorption maxima at 206 nm and 225 nm and can therefore be detected using a UV absorbance detector [195,196]. They have emphasized that the separation of dithionite from other sulfur anions by ion-pair chromatography reported by Weiss and Göbl [191] is in error as the peak assigned to dithionite was in fact due to sulfate.

For the analysis of solid samples of dithionite spanning 50 years, James and coworkers have used iodometric titration, IC as well as Raman spectroscopy [193]. As is known that Raman spectrum of dithionite differs sharply from that of the solid ion [133,197], and the system is also complicated by the spontaneous decomposition of dithionite, James and coworkers have performed analysis of the aqueous samples just to determine if any substantial changes occur during the solution preparation required for the titration and IC analysis. Raman spectroscopy provided qualitative information on the species present in each sample.

A rapid and quantitative spectrophotometric procedure for the determination of dithionite based on the reduction of naphthol yellow S in ammonia containing solution and measurement of the absorbance of the product formed at 502 nm was described by de Carvalho and Schwedt [198]. The results were shown to be in good agreement with those of polarographic determinations. Concentration of dithionite in aqueous solution can also be determined by direct UV spectroscopic analysis (λ_{max} = 315 nm, ε_{max} = 8043 ± 21 $M^{-1}cm^{-1}$) [58].

Sodium hydroxymethanesulfinate (rongalite, HMS) is commercially available as the dihydrate $HOCH_2SO_2Na \cdot 2H_2O$, which crystallizes as white needles, m.p. 63°C [36]. Its solubility in water at room temperature is 60/100 g. The salt is sparingly soluble in alcohol. The water of crystallization is lost upon heating to 120°C. At 140°C the compound undergoes a very exothermic decomposition to formaldehyde, hydrogen sulfide and sulfate [157,200]:

$$2HOCH_2SO_2Na \longrightarrow 2CH_2O + H_2S + Na_2SO_4. \qquad (4.4)$$

The LD_{50} value of sodium hydroxymethanesulfinate is ca. 6400 mg/kg (oral, rat) [36]. The content of HMS can be determined iodometrically as well as by alternate methods which are analogous to determination of the content of dithionite described above [184,189,193,195,199,201]. A rapid differentiation of dithionite and hydroxymethanesulfinate is based on the fact that dithionite reacts immediately with alcoholic ammoniacal solutions of o- or p-dinitrobenzene to give a violet or orange color, respectively, whereas HMS reacts either slowly or not at all [200]. In contrast to dithionite, thermal decomposition of HMS is accompanied by formation of hydrogen sulfide (see Eq. (4.4)). Therefore, HMS may be detected through the hydrogen sulfide it yields on pyrolysis [200].

Thiourea dioxide $(NH_2)_2CSO_2$ (TDO) is a white crystalline compound, m.p. 128°C (decomp.). It is sparingly soluble in water and other common solvents. Thus, the solubility of TDO at 20°C in water is approximately 3%, but increases rapidly with rise in the concentrations of alkali and temperature [78]. The compound dissolves in alkaline solutions with decomposition, but it can be heated to 100°C in concentrated sulfuric acid without decomposition [36]. Decomposition of solid TDO occurs at 120°C and is accompanied by formation of urea, sulfur and sulfur dioxide [46,157,202,203]:

$$2(NH_2)_2CSO_2 \longrightarrow 2(NH_2)_2CO + S + SO_2. \qquad (4.5)$$

Solid thiourea trioxide decomposes according to the following equation [202]:

$$(NH_2)_2CSO_3 \longrightarrow (NH_2)_2CO + SO_2. \tag{4.6}$$

The LD_{50} value of TDO is ca. 1120 mg/kg (oral, rat), and this compound is a skin irritant [36].

Since in acidic solutions reaction of thiourea oxides with iodine at comparable concentrations proceeds quite slowly and depends on the aging of thiourea oxides solutions [204,205] (in the case of trioxide) iodometric titration does not give reliable results. Therefore, for determination of TDO and thiourea trioxide different physico-chemical methods are used. Thus, for the quantitative determination of TDO and thiourea trioxide Gao and coworkers have employed reversed-phase ion-pair high-performance liquid chromatography [206]. The HPLC technique makes it possible to monitor simultaneously concentrations of TDO, thiourea trioxide, thiourea and formamidine disulfide. Concentration of TDO, N-methylthiourea and N,N$'$-dimethylthiourea dioxides in aqueous solutions may be determined by UV spectroscopy ($\lambda_{max} = 270$ nm, $\varepsilon = 489$ M^{-1}cm^{-1}; $\lambda_{max} = 270$ nm, $\varepsilon = 521$ M^{-1}cm^{-1}; $\lambda_{max} = 263$ nm, $\varepsilon = 475$ M^{-1}cm^{-1}, respectively) [207]. Thiourea trioxide absorbs at shorter wavelengths ($\lambda_{max} = 202$ nm, $\varepsilon = 7295$ M^{-1}cm^{-1}) [155]. Other methods such as ^{1}H and ^{13}C NMR as well as IR spectroscopy may also be used for determination of thiourea oxides [89,95,97,155,208].

Chapter 5

Stability in Solutions under Anaerobic and Aerobic Conditions

One of the important features of redox reactions of sulfur-containing reductants is that many times their reducing capability is not connected directly to the original sulfur-containing compound but rather to their decomposition products in solutions. It is well known that, for example, sulfoxylate ion or sulfur dioxide anion radical has much stronger reducing ability than the corresponding thiourea dioxide (TDO) or dithionite ion. As a result, decomposition of these reductants and their reactions with oxidants are closely interrelated. The stability of these compounds in aqueous solutions depends appreciably on the pH and on the presence of dissolved oxygen. Sodium dithionite and hydroxymethanesulfinate are less stable in acidic media, but in contrast to this TDO is unstable in alkaline solutions.

Anaerobic decomposition of sodium dithionite in acidic solutions have been studied in detail [173,182,196,209–221]. The most widely used method for determination of intermediate and final products of its decomposition in solutions is polarography. This method allows to determine simultaneously concentrations of dithionite, sulfide, sulfur dioxide, thiosulfate and active sulfur, S_a (this term implies atomic sulfur, hydrate $S \cdot H_2O$, sulfenic acid (thioperoxide) HSOH, polymeric diradical, sulfur in the polysulfidic chain or in sulfanemonosulfonic acids [182]). It should be noted, however, that formation of a singlet sulfur atom is not a reasonable pathway for the loss of sulfur to occur [222,223]; the energy of formation of a sulfur atom S ($_1D^2$) has been determined to be 277.4 kJ $\cdot$ mol^{-1}. Owing to this high energy, single sulfur atom decomposition pathways cannot account for many of observed sulfur extrusions that occur under mild conditions; the sulfur loss must proceed through a lower energy transient species such as diatomic sulfur (S_2) or longer chain fragment.

Decomposition of dithionite in acidic solutions was found to be catalyzed by active sulfur and sulfide [182]. It was concluded that the following reactions are mainly responsible for the dithionite decomposition:

- Non-catalyzed reaction proceeding during the induction period:

$$2H_2S_2O_4 \longrightarrow S_a + 3SO_2 + 2H_2O. \tag{5.1}$$

- Autocatalytic reaction mainly proceeding at the fast decomposition stage:

$$3H_2S_2O_4 \longrightarrow H_2S + 5SO_2 + 2H_2O. \tag{5.2}$$

Calculations on the basis of proposed mathematical model [217] demonstrated that mentioned two-step scheme adequately describes the process of decomposition.

Holman and Bennett have studied decomposition of sodium dithionite in weakly acidic solutions in the presence of additives of bisulfite [219]. Multivariate analysis of attenuated total reflectance Fourier transform infrared spectroscopy (FTIR) [224] used by them provided multicomponent kinetic data at temperatures from 42°C to 88.5°C. Virtually all the sulfurs in these solutions were accounted for by measuring the quantities of seven known IR-active anions (dithionite, thiosulfate, bisulfite, trithionate, metabisulfite, sulfite and sulfate). There were no additional spectral features observed in the S–O stretching region (750–1350 cm^{-1}). It was shown that addition of trithionate $S_3O_6^{2-}$ substantially accelerates decomposition process under high bisulfite conditions. This ion was also observed as a final product. Holman and Bennett proposed two mechanisms (heterolytic and homolytic) for decomposition of dithionite in the presence of high quantities of bisulfite:

$$S_2O_4^{2-} + HSO_3^- \longrightarrow [O_2S(O_2)S - SO_2OH]^{3-}, \tag{5.3}$$

$$[O_2S(O_2)S - SO_2OH]^{3-} + H^+ \longrightarrow S_3O_6^{2-} + H_2O, \tag{5.4}$$

where Eq. (5.3) is the rate determining step. Trithionate $S_3O_6^{2-}$ participates also in the homolytic pathway:

$$S_2O_4^{2-} \rightleftharpoons 2 \cdot SO_2^-, \tag{5.5}$$

$$HSO_3^- + \cdot SO_2^- \rightleftharpoons \cdot S_2O_5H^{2-}, \tag{5.6}$$

$$\cdot S_2O_5H^{2-} + S_3O_6^{2-} \longrightarrow SO_2 + HSO_3^- + SO_3^{2-} + \cdot S_2O_3^-, \tag{5.7}$$

$$\cdot S_2O_3^- + \cdot SO_2^- \longrightarrow S_2O_3^{2-} + SO_2. \tag{5.8}$$

Reaction between two dianions (Eq. (5.7)) is the rate determining step.

At low concentrations of sulfite the process proceeds via a third pathway:

$$S_2O_4^{2-} + HSO_3^- \longrightarrow [O_2S(O_2)S - OSOOH]^{3-}, \tag{5.9}$$

$$[O_2S(O_2)S - OSOOH]^{3-} \longrightarrow S_2O_5^{2-} + HSO_2^-, \tag{5.10}$$

$$2HSO_2^- \longrightarrow S_2O_3^{2-} + H_2O, \tag{5.11}$$

$$S_2O_5^{2-} + H_2O \rightleftharpoons 2HSO_3^-. \tag{5.12}$$

This pathway precedes and follows a very rapid, apparently autocatalytic decomposition. Note that, in contrast to Eq. (5.3), product of reaction between dithionite and bisulfite (Eq. (5.9)) does not have additional S–S bond what promotes the formation of $S_2O_5^{2-}$ but not that of trithionate as in the first pathway. The serious drawback of the mechanisms of Holman and Bennett is that nothing is said about the role of sulfur and sulfide in the dithionite decomposition. Indeed, at high concentration of sulfite they do not seem to play a significant role since they can rapidly react with bisulfite in acidic media, but in the case of low concentration of sulfite their role cannot be neglected. In addition, catalytic effect of trithionate can be explained by direct formation of sulfur (see Eq. (5.13)) or by its formation from intermediate product of trithionate decomposition — thiosulfate [225], but not by the influence of trithionate itself. The other surprising point is why in the first heterolytic mechanism authors do not consider reaction of dithionite with metabisulfite $S_2O_5^{2-}$ though at high $[HSO_3^-]$ its concentration is significant (see Eq. (5.12)) [226].

Data of Holman and Bennett have been used by Kovács and Rábai for elaboration of their model of oscillatory decomposition of dithionite in a continuous-flow stirred tank reactor (CSTR) [221]. It has been earlier reported that during the decomposition, at elevated temperatures, the concentration of dithionite ions and the pH show several small peaks in time, even in a thermodynamically closed system [215]. DePoy and Mason have suggested mechanisms which could produce the type of oscillations which have been observed [227].

However, observations of Rinker *et al.* [215] have never been confirmed by other laboratories. Numerous efforts of Kovács and Rábai failed to reproduce pH or redox potential oscillations in a closed reactor [220]. Based on these results, they concluded that the thermal decomposition of aqueous dithionite does not exhibit concentration oscillations in a closed reactor. A year later they carried out experiments in a CSTR where the reaction is kept far from equilibrium by the continuous flow and observed large amplitude-sustained pH-oscillations [221]. Oscillatory behavior could be observed in a rather wide range of experimental constraints, i.e. the temperature (25–60°C), the input concentration of $S_2O_4^{2-}$ and the flow rate. To explain experimental observations, Kovács and Rábai suggested a mechanism consisting of three protonation equilibria and seven redox reactions between sulfur species [221]. They assumed that the autocatalytic route is moderated by reaction (Eq. (5.13)), which removes trithionate $S_3O_6^{2-}$ from the autocatalytic cycle by transferring it to the much less reactive $S_2O_6^{2-}$:

$$S_3O_6^{2-} \longrightarrow S_2O_6^{2-} + S. \qquad (5.13)$$

The important negative feedback process is also Eq. (5.14) because it removes bisulfite from the autocatalytic cycle:

$$HSO_3^- + S_2O_3^{2-} \longrightarrow SO_4^{2-} + S + HSO_2^-. \qquad (5.14)$$

The governing role of H^+ is also reflected by this mechanism. Since the protonated $HS_2O_4^-$ reacts much faster with HSO_3^- than $S_2O_4^{2-}$ does, the overall process accelerates with increasing H^+ concentration.

Kovács and Rábai have emphasized that their data demonstrate the first example of oscillations in the system consisting of a single compound (usually oscillatory chemical reaction systems consist of two or more reactants) and can help to improve the general model of sulfur-based oscillators suggested by Rushing and coworkers [228]. It should be noted, however, that though experimental data described by Kovács and Rábai are very interesting, mechanism suggested for their explanation includes some unlikely reactions. For example, reaction between bisulfite and thiosulfate (Eq. (5.14)) assumes formation of sulfoxylate what is very unlikely (otherwise mixture of bisulfite and thiosulfate should possess very strong reducing properties but it is not the case). Kovács and Rábai have observed pH oscillations in the pH range approximately 5.0–6.4 [221], but at such pHs dithionite exists completely in the deprotonated form ($S_2O_4^{2-}$) [229].

Anaerobic decomposition of alkaline solution of sodium dithionite has been much less studied than in acidic conditions due to its high stability at high pH [199,230]. The kinetics of the anaerobic decomposition of dithionite in alkaline solution has been reported to be first order with respect to dithionite [230]. The rate increased very slightly with an increase in alkali at low hydroxide concentrations. The rate was not observed to be proportional to alkali at higher alkali concentration. It was found that a pH region of maximum stability exists [199]. In contrast to Lister and Garvie's data [230], the results of Kilroy [199] show that the half life of dithionite during decomposition is a function of the initial dithionite concentration. In addition, for equimolar alkali to dithionite solutions, sufficient hydroxide was consumed during the decomposition to drastically decrease the pH of the solution. Importantly, below a critical pH region, the dithionite decomposition accelerates exhibiting a characteristic of autocatalytic behavior [199]. Besides the main final products of dithionite decomposition (thiosulfate and sulfite), Kilroy have observed the transient formation of sulfide. After complete decomposition, no sulfide was present. In his paper [199], Kilroy did not determine species to be responsible for the autocatalysis. By analogy with acidic decomposition [182], sulfide can be the autocatalyst. Indeed, increasing the relative proportion of alkali to dithionite greatly enhances the rate of decomposition and, at the same time, concentrations of sulfide.

Decomposition of sodium hydroxymethanesulfinate in acidic solutions resembles to that of dithionite [231]. It is autocatalytic, being accelerated by active sulfur. This effect has been proven by addition of different quantities of thiosulfate which produces sulfur in acidic solutions (Münchow and Steudel [196] have observed the accelerating effect of thiosulfate in the decomposition of dithionite at pH < 6). The addition of sulfite leads to the formation of dithionite due to the following reactions:

$$HOCH_2SO_2^- \rightleftharpoons SO_2H^- + CH_2O, \qquad (5.15)$$

$$HSO_3^- + CH_2O \longrightarrow HOCH_2SO_3^-, \qquad (5.16)$$

$$SO_2H^- + HSO_3^- \rightleftharpoons S_2O_4^{2-} + H_2O. \qquad (5.17)$$

TDO is very stable in acidic solutions and decomposes very slowly in acidic solutions at room temperature. On heating in glacial acetic acid TDO and some of its analogs decompose to give formamidine acetate and sulfur dioxide [207].

Decomposition of thiourea oxides in alkaline solutions has been studied by many investigators, but, in contrast to dithionite, these studies were dedicated mostly for determination of final products (almost without mechanistic details) or comparison of stability of different thiourea oxides. Unfortunately, in many publications no attention is paid to the role of oxygen on the decomposition of TDO and other oxides (influence of oxygen will be discussed later). Often it is impossible to grasp, whether the reaction was carried out in inert atmosphere or in the presence of oxygen (air). Therefore, if no special remarks has been made, we will assume that the experiments have been performed in air atmosphere (in the case of dithionite situation it is much easier since it rapidly reacts with O_2, see later). In contrast to sulfur-containing products, transformation of nitrogen-containing part of TDO molecule does not depend on the presence of air oxygen, therefore experimental circumstances is not a central part to be considered on the final composition of N-containing products.

The corresponding ureas are the only nitrogen-containing products of thiourea, N-methylthiourea and N,N-dimethylthiourea dioxides in strongly alkaline solutions reported by Svarovsky and his coworkers [207]. They have shown that in strongly alkaline solutions dioxide of N,N-dimethylthiourea is more unstable, than TDO, and the most stable is the dioxide of N-methylthiourea. In less alkaline media ($pH \leq 10$) cyanamides as well as ammonia are also formed [39]. Under similar conditions ($pH \approx 10$), thiourea trioxide gives melamine as well as cyanoguanidine [39]. In weakly acidic unbuffered solutions, thiourea trioxide (TTO) produces ammonia and rate of its formation is equal to the rate of TTO decomposition [155]. In addition to the signal of ammonia, during decomposition of TTO in ^{1}H NMR spectra in D_2O the other signal was observed, but it was not assigned to any compound. In contrast to dioxide, decomposition of TTO at pH around 13–14 is accompanied mainly by formation of cyanamide [39].

In acidic solutions ($pH = 3–7$), TDO is more stable than TTO [232]. But at $pH > 7$ stability of TTO is higher than that of TDO. These observations can be explained by differences in acid–base properties of dioxide and trioxide. pK of TDO can be estimated from the data on rate constants of decomposition [207,232] ($pK \approx 8.5$ at 298 K). A value of pK of TTO has not been published, so it is impossible to compare pKs of dioxide and trioxide.

Numerous papers are dedicated to interaction of sodium dithionite with oxygen [209,233–250]. This can be explained by potent oxygen-scavenging properties of dithionite. Owing to them it is used for oxygen removal from closed water systems [251].

The first air oxidation study of dithionite was conducted by Meyer [209]. He has shown that the products of reaction were sulfite and sulfate. Later Nicloux has also determined that in reaction with molecular oxygen an equimolar mixture of sulfite and sulfate was formed [234]. The first kinetic study of this reaction has been performed by Lynn [235]. He has determined that oxidation of $Na_2S_2O_4$ proceeded according to a first-order mechanism with respect to dithionite. But 6 years later, Rinker and coworkers have shown that oxidation in 0.1 M NaOH was half and first order with respect to dithionite and oxygen, respectively [236]. The half-order mechanism was attributable to the presence of sulfur dioxide anion radical as an intermediate, i.e. the reaction obeyed the following rate law:

$$-\frac{d[S_2O_4^{2-}]}{dt} = k_1[O_2][S_2O_4^{2-}]^{0.5}, \qquad (5.18)$$

with $k_1 = 0.15$ M$^{0.5}$s^{-1} at 30°C in 0.1 M NaOH.

They invoked the following scheme, where the overall process started with Eq. (5.5) assuming an equilibrium constant of $K_2 = k_2/k_{-2} = 10^{-9}$ M [236] followed by Eq. (5.19) being the rate determining step to rationalize the rate law indicated above:

$$\cdot SO_2^- + O_2 \xrightarrow{k_3} \text{products}, \qquad (5.19)$$

where k_3 was found to be 4000 M^{-1}s^{-1} at 30°C [237].

The rate of removal of oxygen from aqueous solution by sodium dithionite in 0.1 M sodium hydroxide was also studied by Morello and coworkers [239]. The measurements were made at 37°C, so that the data would be applicable in studies of the kinetics of oxyhemoglobin in blood. The reaction over the first 40 ms was found to be first order with respect to dithionite and zero order with respect to molecular oxygen. The initial rate constant was determined to be 42.5 ± 3.6 s^{-1}. This result is inconsistent with Eq. (5.18). But, if the rate of production of $\cdot SO_2^-$ anion radicals (Eq. (5.5)) is the rate determining, it becomes consistent with the above scheme (Eqs. (5.5) and (5.19)).

In 1974, Creutz and Sutin [238] have reported the results of the studies of the oxygen–dithionite reaction which differ significantly from the one mentioned before. When O_2 was in excess, the disappearance of dithionite was exponential, giving excellent first-order fits. Dissociation of dithionite anion is the rate determining step under these conditions (at 25°C $k_2 = 2.5$ s^{-1} and 1.8 s^{-1} at pH = 6.5 and in 0.1 M NaOH, respectively, in excellent agreement with value measured by Lambeth and Palmer [237]). With dithionite as a reagent in excess, zero-order decays were observed. As mentioned earlier, the zero-order decay is observed if Eq. (5.5) is rate-determining, i.e. when

$$k_3[O_2][\cdot SO_2^-] > k_{-2}[\cdot SO_2^-]^2. \qquad (5.20)$$

From the above-mentioned considerations the limit $k_3 \geq 10^8$ M^{-1}s^{-1} was found, which is much higher than those proposed in previous studies. Later, Huie and coworkers [252] have determined a value of this rate constant $(2.4 \times 10^9$ M^{-1}s$^{-1})$. Authors [238] pointed out that the difference between their and Rinker's data [236] may have resulted from insufficient stirring used in the previous study, i.e. if the steady state O_2 level is far below that of saturation, Eq. (5.19) becomes rate determining.

Tao and coworkers have found that stoichiometry of reaction between dithionite and oxygen depends also on the temperature [240]. Thus, the ratio of dithionite to oxygen consumed in the reaction is 1.2 ± 0.2 at 25°C and 1.7 ± 0.1 at 37°C. This result shows that at higher temperature the input of the following reactions is larger. An activation energy of 17.5 kcal/mol (73.2 kJ/mol) for dithionite–oxygen reaction has been calculated by Singh and coworkers [241]. Similar result (76.2 kJ/mol in the temperature range 15–34°C) was received by Kawagoe *et al.* [242]. Both groups have reported first order with respect to dithionite and zero order with respect to oxygen [241,242].

Stoichiometry of dithionite/oxygen reaction depends also on $[S_2O_4^{2-}]$ [238]. In excess of dithionite a rapid zero-order decay appears always to consume 1 mol $S_2O_4^{2-}$ per 1 mol O_2. At 2 mM $S_2O_4^{2-}$ no further reaction occurs. At 0.2 mM $S_2O_4^{2-}$ approximately 2 mols of dithionite per one mole of oxygen are consumed due to slow subsequent reactions. The nature of these reactions is, however, obscure [238]. In principle, it could be reactions with reduced oxygen species, for example, hydrogen peroxide. Creutz and Sutin have studied the reaction between sodium dithionite and H_2O_2 (or HO_2^-) under the same conditions as with oxygen [238]. The disappearance of dithionite was half order with respect to dithionite. This result showed, that reaction of $\cdot SO_2^-$ with H_2O_2 (HO_2^-) is rate determining and it is much slower than reaction of sulfur dioxide anion radical with oxygen. Indeed, second-order rate constants of reactions of $\cdot SO_2^-$ with H_2O_2 and HO_2^- are 260 and 5.9 $M^{-1}s^{-1}$, respectively. Thus, depending on the system and conditions used, either the dissociation of dithionite or the reaction of the $\cdot SO_2^-$ with oxidant can be the rate determining process.

Calorimetric studies of the reduction of oxygen in solution by sodium dithionite are also in agreement with a stoichiometry of two moles $Na_2S_2O_4$ per mole of oxygen [246]. The reaction is biphasic with $\Delta H_T = -118 \pm 7$ kcal/mol (-494 ± 29 kJ/mol). The reduction of hydrogen peroxide by dithionite in 0.1 M phosphate buffer, at pH = 7.15, is a much slower process and with an enthalpy change of ca. -74 kcal/mol (-314 kJ/mol).

Next study of reaction between sodium dithionite and hydrogen peroxide has been performed by Kovács and Rábai [220]. The kinetics of the reaction between $S_2O_4^{2-}$ and H_2O_2 is found to be very complex both in a closed reactor and in a CSTR in an unbuffered aqueous solution. The main products of the oxidation are SO_4^{2-} and $S_2O_6^{2-}$ in excess of H_2O_2. The measured pH-time traces show multiple inflection points in a closed reactor, suggesting that the reaction takes place in several distinct steps. The pH exhibits very large amplitude relaxation oscillations in a CSTR in a narrow range of input concentrations, flow rate and temperature. A simple empirical rate law model consisting of three redox reactions and three protonation equilibria is proposed:

$$S_2O_4^{2-} + H_2O_2 \longrightarrow 2HSO_3^-, \tag{5.21}$$

$$HSO_3^- + H_2O_2 \longrightarrow SO_4^{2-} + H^+ + H_2O, \tag{5.22}$$

$$2H_2O_2 + S_2O_4^{2-} + H^+ \longrightarrow HS_2O_6^- + 2H_2O, \tag{5.23}$$

$$H^+ + SO_3^- \rightleftharpoons HSO_3^-, \tag{5.24}$$

$$H^+ + S_2O_6^{2-} \rightleftharpoons HS_2O_6^-, \tag{5.25}$$

$$H^+ + OH^- \rightleftharpoons H_2O. \tag{5.26}$$

A series of papers have been dedicated to absorption of oxygen by sodium dithionite solutions [243–250]. The data received by different groups are contradictory mainly due to (a) dithionite decomposition (when pH < 8) taking place when the starting alkaline solution is not sufficiently concentrated or when the pH is not kept constant by the addition of base, (b) absorption does not occur in the fast reaction regime. Camacho *et al.* have shown that absorption occurs by means of a fast reaction of zero order with respect to oxygen and 1.5 order with respect to dithionite [247]. A mechanism with intermediate formation of sulfate radical (in direct reaction of $\cdot SO_2^-$ and oxygen) as well as formation of hydroxyl radical is proposed to explain these results. In the opinion of Tao and coworkers [240], the reaction of $\cdot SO_2^-$ with O_2 very probably produces SO_2 and superoxide, and SO_2 in aqueous solution already yields HSO_3^-. Subsequent reactions involve further oxidations by O_2 and $\cdot O_2^-$.

A detailed kinetic and mechanistic study of aerobic decomposition of sodium hydroxymethanesulfinate has been performed by Makarov *et al* [253]. They have shown that in the course of HMS decomposition under anaerobic and aerobic conditions dithionite is formed, but in the second case induction period is observed. This period persists as long as molecular oxygen is present in the reaction solution. The complete consumption of molecular oxygen is a prerequisite for the formation of $S_2O_4^{2-}$. Added sulfite dramatically increases the rate of dithionite formation in HMS solutions. Therefore, it was assumed that dithionite is formed in the reaction between sulfoxylate (primary product of HMS decomposition) and bisulfite (Eq. (5.17)).

The mechanism of dithionite formation in TDO solutions is absolutely different. In this difference pH plays a crucial role. Svarovsky and coworkers observed that decomposition of TDO in air-saturated alkaline solutions is accompanied by formation of dithionite, preceded by an induction period [207]. But without oxygen dithionite does not appear at all. Dithionite was also formed in the presence of KO_2 and H_2O_2 under anaerobic conditions. Additives of sulfite did not influence the rate of dithionite formation. Authors have concluded that under aerobic conditions dithionite is formed not due to reaction between sulfoxylate and sulfite but due to the following reactions:

$$SO_2^{2-} + O_2 \longrightarrow \bullet SO_2^- + \bullet O_2^-, \tag{5.27}$$

$$\bullet SO_2^- + O_2 \longrightarrow SO_2 + \bullet O_2^-, \tag{5.28}$$

$$SO_2^{2-} + \bullet O_2^- \longrightarrow \bullet SO_2^- + O_2^{2-}, \tag{5.29}$$

$$\bullet SO_2^- + \bullet O_2^- \longrightarrow SO_2 + O_2^{2-}, \tag{5.30}$$

$$SO_2^{2-} + O_2^{2-} + 2H_2O \longrightarrow \bullet SO_2^- + 3OH^- + \bullet OH, \tag{5.31}$$

$$\bullet SO_2^- + O_2^{2-} + 2H_2O \longrightarrow SO_2 + 3OH^- + \bullet OH. \tag{5.32}$$

This sequence of reaction has to be incorporated as well by Eq. (5.5). Their data unequivocally proved the heterolytic mechanism of decomposition of TDO and formation of sulfoxylate in its alkaline solutions.

Data mentioned above show that sulfoxylate and sulfur dioxide anion radical play a major role in the decomposition of dithionite, hydroxymethanesulfinate and TDO. Properties of sulfoxylate have been recently considered in detail in the review [44], therefore we will not describe them here. But sulfur dioxide anion radical deserves a separate consideration since there is no recent review on its chemistry, and a significant role of sulfur dioxide as well as its oxidized and reduced forms in biology has been recognized [254,255]. Recent studies showed that SO_2 can be generated endogenously in mammals, and its protective effects have been found. Endogeneous SO_2 has antioxidant, anti-inflammatory, anti-hypertension and anti-atherogenic effects and regulates vascular tone and cardiac function in mammals [255]. These discoveries suggest a novel role of endogenous SO_2 in the modulation of the cardiovascular system and provide a basis for new treatments for cardiovascular diseases [254]. Although the biological oxidation of SO_2 (bisulfite) is certainly more common than its reduction [256], under anaerobic conditions sulfur dioxide (bisulfite) is reduced to sulfur dioxide anion radical (dithionite) by either reduced flavodoxins or a mixture of paraquat, H_2, and hydrogenase. Sulfur dioxide anion radical is also an intermediate of redox reactions of the other important signal molecule — hydrogen sulfide [257], being a product of reaction between •HS and O_2 [258].

The chemistry of sulfur dioxide anion radical began in 1950s when three pioneer papers on sodium dithionite were published [129,259,260]. That time, •SO_2^- was detected by electron paramagnetic resonance (EPR) though the existence of this species was suggested earlier to account for the polarographic behavior of dithionite in aqueous solutions [180,218] (note, for clarity, that Furness wrote not about •SO_2^-, but unknown second form of dithionite whose concentration increases at higher temperatures [180]).

In 1961, Clark and coworkers after X-irradiation of solid sodium and potassium dithionite have received paramagnetic species having an ultraviolet absorption band in the 350 nm region. This was tentatively identified as the radical-ion •SO_2^-. Electron-spin resonance spectra of the irradiated solids are closely similar to that obtained from sodium dithionite moistened with water, having a g-value of 2.004 and a half-width of 11.5 gauss [261]. Further analysis of EPR spectrum of •SO_2^- were done by Atkins and coworkers [262]. Mayhew has calculated midpoint redox potential for the couple •SO_2^-/HSO_3^- at pH = 7 and 25°C (-0.66 V) [263]. Neta and coworkers have estimated the self-exchange rate for the couple $SO_2/$•SO_2^- in acidic solutions [264]. The calculated values were found to vary over many orders of magnitude, similar to the situation reported before for the $O_2/$•O_2^- couple. In the authors' opinion, such variations may be a general characteristic of redox pairs composed of a small number of atoms and one form of which is neutral, so that solvation of the two species varies greatly.

SO_2 was reported to be reduced rapidly by •CO_2^- to produce •SO_2^- while HSO_3^- and SO_3^{2-}, were unreactive [265]. Also Ti^{3+} was found to react with sulfite in acid solutions (pH = 2–6) to produce •SO_2^- [266].

The •SO_2^- reduces the aromatic nitro compounds to the corresponding anion radicals, but did not abstract the hydrogen from the saturated compounds as well as it is not added to the unsaturated compounds [267]. Harrington and Wilkins have compared reactivity of different reductants towards methemerythrin and received the following sequence: e_{aq}^- > •CO_2^- > •SO_2^- (from dithionite reduction) [268]. This sequence is the same as that observed with reduction of heme proteins but the rate constants are some 10–100 times smaller for the former one.

It has been claimed that both $\cdot SO_2^-$ and $\cdot SO_3^-$ can be spin trapped by 2-methyl-2-nitrosopropane (MNP) to yield spin adducts with indistinguishible hyperfine coupling constants [269–271]. But it was shown later that sulfur dioxide anion radicals does not form persistent spin adducts with spin traps (nitroso compounds and nitrones) [272,273]. No spin adducts were also detected with 5,5-dimethylpyrroline-1-oxide (DMPO) although the latter is known to form persistent spin adducts with $\cdot SO_3^-$ [274].

A doping of $\cdot SO_2^-$ into $SrTiO_3$, or $SrZrO_3$ has been performed by Yonemura *et al.* [275]. The heating of $SrTiO_3$, or $SrZrO_3$, alternately in CS_2 (800°C) and O_2 (600°C) has been demonstrated as a method for the introduction of $\cdot SO_2^-$ dopant ions. When this heating cycle is repeated several times, a paramagnetic species having g-factor components of 2.0069, 2.0115 and 2.0021 is generated in these compounds. This species is identified as $\cdot SO_2^-$. The formation of $\cdot SO_2^-$ in these wide band gap semiconductors adds a visible light response to them.

The ability of dithionite to give EPR signal should be taken into account in the interpretation of results received with using of this compound as a reductant. $Na_2S_2O_4$ is often used in studies of redox proteins [276]. Yu and coworkers found [276] that when ascorbate/Cytochrome (Cyt c) or dithionite are used to reduce bovine Cytochrome c Oxidase (CcO) prior to its reaction with oxygen, ascorbyl or $\cdot SO_2^-$ radicals are generated and may be trapped under conditions typically used to detect reaction intermediates. The $\cdot SO_2^-$ anion radical has rhombic symmetry with g-values of $g_x = 2.0089$, $g_y = 2.0052$ and $g_z = 2.0017$. When the contributions from the ascorbyl and $\cdot SO_2^-$ radicals were removed, no protein-based radical on CcO could be identified in the EPR spectra.

Interestingly, that $\cdot SO_2^-$ anion radical is detected not only in the solid sodium dithionite and during its decomposition, but also in the course of thermal decomposition of sodium and potassium metabisulfite, and sodium bisulfite [277]. A possibility for metabisulfites as in case of dithionite is radical transmission through the lattice which would effectively isolate the radicals from each other, e.g.

$$[^-O_2S\cdot \ \cdot SO_3^- \ ^-O_2SSO_3^-] \longrightarrow [^-O_2S\cdot \ SO_3^{2-} \ O_2S \ \cdot SO_3^-] \longrightarrow \text{etc.}$$

$$(5.33)$$

The detection of $\cdot SO_2^-$ in the thermal decomposition of $NaHSO_3$ is perhaps the most surprising. Formally, it would appear that a hydroxyl radical should be the other radical produced:

$$HO - SO_2^- \longrightarrow HO\cdot + \cdot SO_2^-.$$

$$(5.34)$$

Reduction of sulfur dioxide in non-aqueous solutions has been studied in detail since SO_2 is involved in several types of batteries [278]. We have already discussed the formation of polythionite prepared by chemical reaction of SO_2 and Li (see Chapter 2). Here we shall consider other features of reduction of sulfur dioxide and nature of the products of this process. By γ-irradiation of sulfur dioxide in 2-methyltetrahydrofuran at $-196°C$, blue glassy solution was obtained [279]. The absorption spectrum of the blue species was identical with that of the stable anion radical produced by the amalgam reduction of sulfur dioxide, which was attributed to the dimer or trimer anion radical. The formation of blue and brown (red) species in the course of chemical or electrochemical reduction of SO_2 has been observed by many other authors [278,280,281]. In his paper, Knittel [280] has reviewed the results of numerous early papers (see Ref. [280]) on the reduction of SO_2 in non-aqueous solutions. He proposed that electrochemical reduction of SO_2 in DMF finally leads to the red species $S_3O_6^{2-}$ via the blue complex

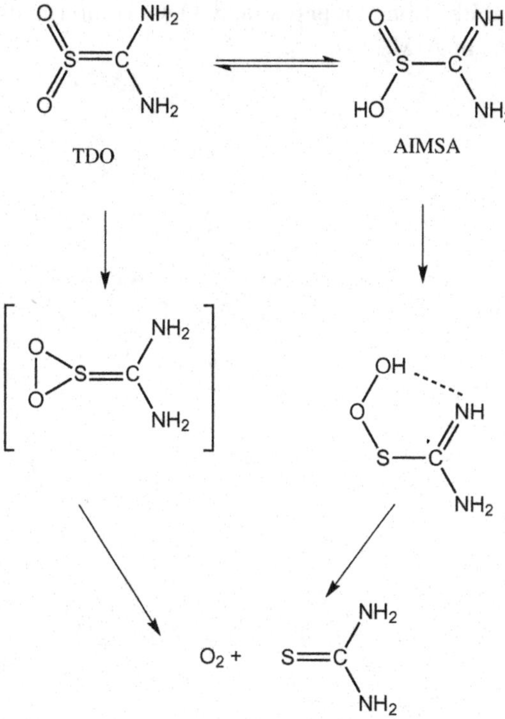

Fig. 5.1   Possible reaction pathways for O_2 release from TDO

$\cdot S_2O_4^-$. However, later Potteau and coworkers have shown [278] that the blue complex is not $\cdot S_2O_4^-$, but its dimer $S_4O_8^{2-}$. In the presence of a supporting electrolyte, the red complex $S_3O_6^{2-}$ is observed and it is shown that this species is formed by reaction between $\cdot SO_2^-$ and $\cdot S_2O_4^-$ (the influence of metal ions on blue and red (brown) species has been studied earlier [281]). It can be seen from the data of Potteau *et al.* that both colored species are *not* radical species.

The most interesting feature of non-aqueous solutions of TDO is a possibility of conversion of TDO to oxygen. Burgess and coworkers have found that TDO decomposes to thiourea, urea (2:3 ratio) and oxygen in refluxing anhydrous acetonitrile with 98% conversion after 0.7 h [282].

The possibility of oxygen liberation from TDO has been confirmed by computational studies [154]. Figure 5.1 shows feasible reaction pathway to release O_2 from TDO. In line with the results shown earlier in this chapter, this solution-phase property of TDO is much easier to be explained by AIMSA tautomer. The significant difference in reactivity between the two tautomers is traceable to the oxygen-bound proton in AIMSA, which allows partial stabilization of a peroxide S–O–O–H intermediate at an O–O distance of 1.5 Å in AIMSA.

Chapter 6

Organic Reactions

6.1 Synthesis of Organofluorine Compounds

Sulfur-containing reductants are used in organofluorine chemistry as sulfina-todehalogenation agents. Sulfinatodehalogenation is one of the most important methods of introducing fluorine atoms in organic molecules [283–285]. It is a simple and efficient reaction for synthesizing of polyfluoroalkane-sulfinates and sulfonates. The reaction can also be applied for the poly-fluoroalkylation of organic compounds [284]. Sulfinatodehalogenation can directly convert a perfluoroalkyl halide into the corresponding per-fluoroalkanesulfinate, without using harsh reaction conditions. Among numerous fluoroalkyl halides, iodides are more reactive than bromides, but chlorides are inert in common conditions. Substrates with long fluoroalkyl chains are usually more reactive than those with short ones [283]. The sulfinatodehalogenation reaction can easily be scaled up to an industrial level. Perfluoroalkanesulfinates can also be transformed into perfluoroalka-nesulfonic acids and their derivatives, which are excellent surfactants and can as well be used to prepare ion exchange membranes [283]. Low price of reagents, mild reaction conditions, good yields and applicability to wide range of substrates all make sulfinatodehalogenation very popular.

The sulfinatodehalogenation reaction was discovered by Huang and his coworkers, in 1981, when the sulfinate salt $-O_2SCF_2CF_2OCF_2CF_2SO_2^-$ (sodium or potassium salt) was formed if the difluoroiodomethyl-containing compound $ICF_2CF_2OCF_2CF_2SO_2F$ was treated with sodium sulfite in aqueous 1,4-dioxane (that is, the CFJ group was transformed into a $CF_2SO_2Na(K)$ group) [286,287]. Interestingly, this reaction does not proceed in aqueous solution. It was found that the presence of a small amount of dioxane hydroperoxide in the dioxane used was responsible for the reaction, i.e. sulfite should be oxidized to sulfite anion radical $\cdot SO_3^-$ [284]. A single electron transfer (SET) process was proposed to explain the course of the process [284]:

$$ROOH \longrightarrow RO\cdot + \cdot OH, \tag{6.1}$$

$$SO_3^{2-} + RO\cdot(\cdot OH) \longrightarrow \cdot SO_3^- + RO^-(OH^-), \tag{6.2}$$

$$R_FI + \cdot SO_3^- \longrightarrow \cdot[R_FISO_3]^- \longrightarrow \cdot R_FISO_2 + IO^-, \tag{6.3}$$

$$\cdot R_FISO_2 + SO_3^{2-} \longrightarrow R_FISO_2^- + \cdot SO_3^-, \tag{6.4}$$

$$SO_3^{2-} + IO^- \longrightarrow SO_4^{2-} + I^-. \tag{6.5}$$

The solvent plays an important role in this process as well. In case of sulfite the most preferable solvents are water–dimethylformamide and water–acetonitrile systems [284].

Later, to improve this reaction, Huang's group suggested more effective reagent — sodium dithionite [288]. From the mid-1980s to early 1990s other reagents have also been studied in different reactions of fluorine-containing compounds including sodium hydroxymethanesulfinate [289–299], zinc hydroxymethanesulfinate [292], thiourea dioxide [300]. Here, sulfoxylate anion plays the role of the reactive species [284]:

$$HOCH_2SO_2^- \rightleftharpoons CH_2O + HSO_2^-, \tag{6.6}$$

$$(NH_2)_2CSO_2 + OH^- \longrightarrow (NH_2)_2CO + HSO_2^-, \tag{6.7}$$

$$R_FI + HSO_2^- \longrightarrow \cdot R_F + I^- + \cdot HSO_2, \tag{6.8}$$

$$\cdot HSO_2 + OH^- \longrightarrow \cdot SO_2^- + H_2O. \tag{6.9}$$

In case of dithionite as well as $Zn-SO_2$ [301,302], the reactive species is sulfur dioxide anion radical $\cdot SO_2^-$ (Fig. 6.1). Nowadays, sodium dithionite remains one of the most frequently used reagent in organofluorine chemistry [293,302–316].

Figure 6.1 shows the intermediary formation of polyfluoroalkyl radicals R_F. A free radical mechanism is supported by trapping R_F by nitroso compounds [283] and olefins [284]. The reactions of R_FI with alkenes and alkynes provide iodide addition products, while reactions of R_FI with aromatics produce substitution products which contain no iodine. R_FBr behaves like R_FI, but the corresponding chlorides do not react in the same way [283]. It was shown that weak electronegativity of halogen and strong negative inductive effects of the perfluoroalkyl group in perfluoroalkyl halides favor sulfinatodehalogenation reaction [283]. The different

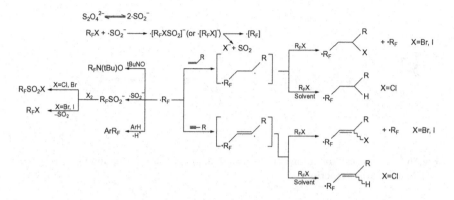

Fig. 6.1 Radical mechanism of sulfinatodehalogenation reaction

examples demonstrating the influence of the structure of perfluoroalkyl bromides and iodides on their reactivity in sulfinatodehalogenation can be found in a recent review [283]. Some chlorocarbons, for example, CCl_4 (but not $HCCl_3$), can be also converted to corresponding sulfinate by its reaction with $Na_2S_2O_4$ [283,317]. Chen and coworkers showed that [283] if dimethylsulfoxide is used instead of CH_3CN/H_2O (as in so-called "standard sulfinatodehalogenation") and temperature is increased from 40–60°C to 80–100°C, the conversion of CF_3CH_2Cl is possible. This reaction was named as "modified sulfinatodehalogenation". To prevent decomposition of sodium dithionite $NaHCO_3$ is often used to keep the solution basic. The reason why DMSO is more suitable for conversion of R_FCl than CH_3CN/H_2O is not yet clear. Possibly, one of the reasons is the influence of water on the equilibrium constant of the following reaction:

$$S_2O_4^{2-} \rightleftharpoons 2 \cdot SO_2^-. \qquad (6.10)$$

Unfortunately, authors of the review [283] do not mention this influence, though they discuss about a possible significant role of water in sulfinatodehalogenation reaction. The practical role of water is to increase solubility of $Na_2S_2O_4$, but simultaneously additives of water strongly decrease equilibrium constant of Eq. (6.10) and make concentration of sulfur dioxide anion radical much lower [63].

DMSO was used as a solvent in the synthesis of various fluorinated extended porphyrins (Fig. 6.2) [308]. The method is based on direct intramolecular cyclization and reductive defluorinative aromatization of β-perfluoroalkylated porphyrins by highly selective C–F bond activation under modified sulfinatodehalogenation reaction conditions. The reaction

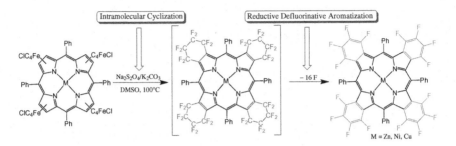

Fig. 6.2 Synthesis of fluorinated extended porphyrins using modified sulfinatodehalogenation reaction

proceeds at sufficiently high temperature (100°C). Indeed, it is known that the increase of temperature facilitates the increase of equilibrium constant of Eq. (6.10) [1].

One of the most important applications of sodium dithionite in organofluorine chemistry is the synthesis of sodium trifluoromethanesulfinate CF_3SO_2Na (Langlois' reagent) [292,307–309,318]. The incorporation of a trifluoromethyl (CF_3) group into organic compounds is of great importance in pharmaceutical, agricultural and material sciences [310]. $CF_3SO_2^-$ is available from CF_3Br and sodium dithionite or Zn/SO_2 [292,308]:

$$\cdot SO_2^- + CF_3Br \longrightarrow SO_2 + \cdot CF_3 + Br^-, \qquad (6.11)$$

$$\cdot CF_3 + \cdot SO_2^- \longrightarrow CF_3SO_2^-. \qquad (6.12)$$

Langlois and coworkers showed [307–309] that trifluoromethyl radicals can be generated from monoelectronic oxidation of sodium trifluoromethanesulfinate, for example, by t-butyl hydroperoxide:

$$CF_3SO_2^- \xrightarrow{-e^-} [\cdot CF_3SO_2] \longrightarrow \cdot CF_3 + SO_2. \qquad (6.13)$$

They carried out synthesis of trifluoromethylated aromatic compounds (monotrifluoromethylation predominates in the presence of catalytic amounts of Cu(II) triflate whereas mono- and bis-trifluoromethylated aromatics are obtained in almost equal quantities in absence of Cu(II)) [307], trifluoromethylated ketones and enol esters (in the presence of catalytic amounts of Cu(II)) [308] and S-trifluoromethyl-containing α-amino acids [309]. In the recent review [310], trifluoromethylation of aryl- or vinylboronic acids and potassium organotrifluoroborates, alkenes, unactivated olefins and aryl(heteroaryl)enol acetates, α,β-unsaturated carboxylic acids, N-arylacrylamides, coumarins, arenes, heterocycles as well as oxy- and hydrotrifluoromethylation of alkenes are described.

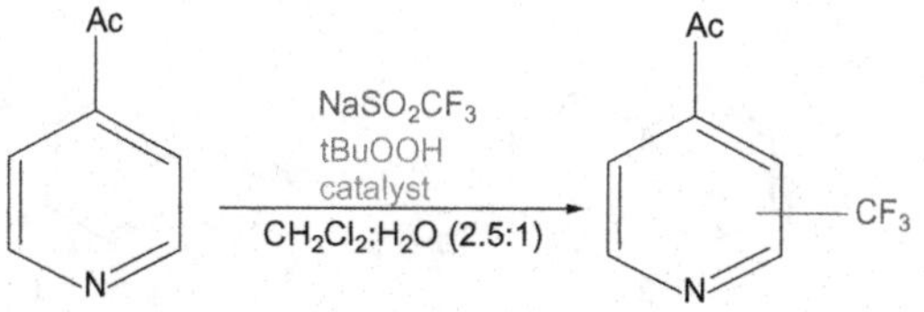

Fig. 6.3   Reaction of 4-acetylpyridine with sodium trifluoromethanesulfinate under oxidative conditions (catalyst — $FeSO_4$, $CuSO_4$ or $CoClO_4$)

In particular, in 2011, it was demonstrated that sodium trifluoromethanesulfinate was capable of trifluoromethylating heterocycles via a radical mechanism [318]. Most of the studies of trifluoromethylation were carried out with 4-acetylpyridine (see Fig. 6.3).

It was shown that metal additives were not required for a productive reaction, although trace metals found in CF_3SO_2Na could be responsible for reaction initiation. Excess reagents were required for appreciable conversion to product. Use of Langlois' reagent gives a possibility to trifluoromethylate pyridines, pyrroles, indoles, pyrimidines, pyrazines, phthalazines, quinoxalines, deazapurine, thiadiazoles, uracils, xanthines and pyrazolino-pyrimidines [318].

Though sodium hydroxymethanesulfinate is used in organofluorine chemistry often in the same reactions and almost under the same conditions as sodium dithionite, sometimes its usage leads to different results. For example, using acetonitrile as cosolvent, perfluoroalkyl iodides **1** reacted at 70–75°C with coumarins **2** in an aqueous solution of sodium hydroxymethanesulfinate in presence of sodium hydrogen carbonate to give 3-perfluoroalkylcoumarins **3** as major products (Fig. 6.4) [296]. The reaction is believed to proceed through a free-radical process. In presence of sodium dithionite, **1** was converted into the corresponding sodium sulfinates completely at room temperature without formation of **3**. When thiourea dioxide was used, compounds **3** were obtained in low yields together with some by-products such as R_FSO_2Na and R_FH. Besides acetonitrile, dimethylformamide and ethanol could also be used as cosolvents, but more R_FH was formed when ethanol was used [296]. It should be noted that a fairly good solubility of sodium hydroxymethanesulfinate in DMF [14] allowed to use this solvent in the reactions of rongalite much more frequently than in the reactions of sodium dithionite. Indeed, HMS was used not only in the synthesis of 3-perfluoroalkylated coumarins, but also 3-perfluoroalkylated thiocoumarines and 2-quionolones (see Fig. 6.5).

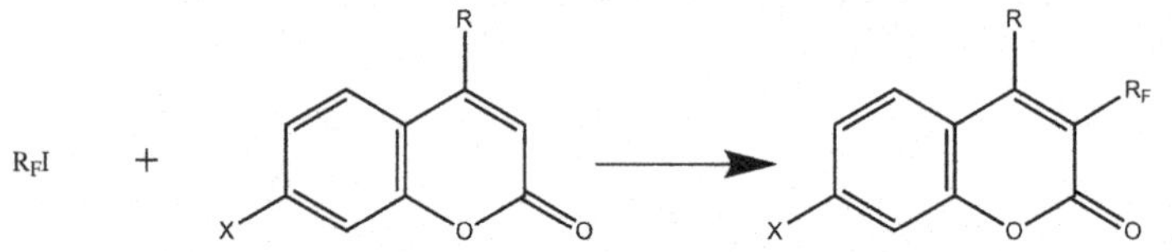

R_FI +

1

a $R_F = F(CF_2)_6$

b $R_F = F(CF_2)_7$

c $R_F = F(CF_2)_8$

d $R_F = Cl(CF_2)_4$

e $R_F = Cl(CF_2)_6$

f $R_F = Cl(CF_2)_8$

2

g $R = X = H$

h $R = Me$, $X = OH$

j $R = Me$, $X = NEt_2$

3

ag $R_F = F(CF_2)_6$, $R = X = H$

bg $R_F = F(CF_2)_7$, $R = X = H$

cg $R_F = F(CF_2)_8$, $R = X = H$

dg $R_F = Cl(CF_2)_4$, $R = X = H$

eg $R_F = Cl(CF_2)_6$, $R = X = H$

fg $R_F = Cl(CF_2)_8$, $R = X = H$

ah $R_F = F(CF_2)_6$, $R = Me$,
    $X = OH$

eh $R_F = Cl(CF_2)_6$, $R = Me$,
    $X = OH$

ej $R_F = Cl(CF_2)_6$, $R = Me$,
    $X = NEt_2$

Fig. 6.4   Reaction of perfluoroalkyl iodides with coumarins

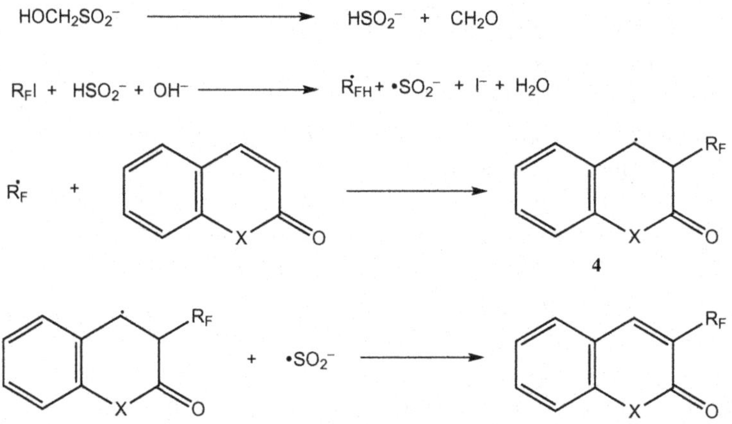

$$HOCH_2SO_2^- \longrightarrow HSO_2^- + CH_2O$$

$$R_FI + HSO_2^- + OH^- \longrightarrow \overset{\bullet}{R}_{FH} + {}^\bullet SO_2^- + I^- + H_2O$$

Fig. 6.5   Synthesis of 3-perfluoroalkylated coumarins ($X = O$), thiocoumarins ($X = S$) and 2-quinolones ($X = NR$) by direct perfluoroalkylation with perfluoroalkyl iodides and sodium hydroxymethanesulfinate

The results obtained from the above reactions may be explained in terms of a radical mechanism indicated in Fig. 6.5

The HSO_2^- anion reacts with R_FI and OH^- to form ${}^\bullet SO_2^-$ and the corresponding ${}^\bullet R_F$ radicals, which then react with coumarins to form a benzylic radical intermediate **4**. Abstraction of hydrogen from **4** by ${}^\bullet SO_2^-$

results in the formation of the title products with the regeneration of HSO_2^- [296].

Thiourea dioxide was introduced as a sulfinatodehalogenating agent almost at the same time as sodium hydroxymethanesulfinate. It was shown that HSO_2^- generated from TDO in aqueous CH_3CN at 35–45°C in presence of sodium bicarbonate reacted with perfluoroalkyl iodides, polyfluoroalkyl bromide, 1,1,1-trichloro-2,2,2-trifluoroethane and carbon tetrachloride to give the corresponding sulfinates in good yield. TDO was also able to initiate the addition of perfluoroalkyl iodide to olefins and alkynes at 30°C [300]. Despite these results, thiourea dioxide is used in organofluorine chemistry much less than sodium dithionite and hydroxymethanesulfinate. After publication of paper [300], besides earlier-mentioned study [296], we found only one article ([293]) where TDO was used in reactions of organofluorine compounds (for conversion of pentafluoroiodobenzene to pentafluorobenzene). Possibly, this can be explained by less solubility of TDO in water and organic solvents. Beside that, as it was shown in the previous chapters, decomposition of TDO in weakly alkaline solutions, for example, in presence of sodium carbonate, is very complex and is accompanied by formation of not only sulfoxylate and urea, but also that of ammonia and cyanamide [39].

6.2 Synthesis of Guanidines

In the syntheses of guanidines nitrogen-containing part of thiourea oxides plays the main role [1]. Guanidine, $HN=C(NH_2)_2$, and substituted guanidines are well known primarily as very strong organic bases ("superbases") [319]. Besides their Brönsted basicity, free guanidines are efficient Lewis bases and hydrogen bond donors and acceptors. On the other hand, guanidinium salts are weak Brönsted acids and bidentate, cationic hydrogen bond donors [320]. Guanidinium-based receptors show high affinities and selectivities for oxyanion substrates (carboxylates, phosphates, sulfates, nitrates) [41]. It was also shown in the last few years that guanidines display high organocatalytic properties [320] and are bioactive compounds [321–325]. The natural amino acid, L-arginine, containing guanidine group (Fig. 6.6), is found in many active sites in proteins and enzymes; it is critical for normal functioning of living organisms [321].

The wide range of activities displayed by guanidines has motivated the development of novel reagents and different synthetic scheme for their preparations. Thioureas and isothioureas are used as common starting

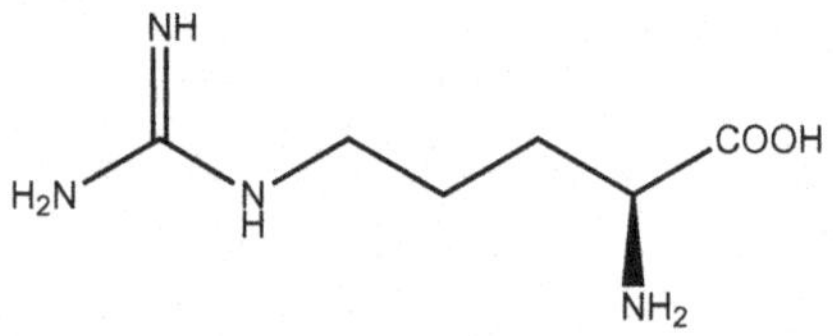

Fig. 6.6 L-Arginine

materials for the synthesis of guanidines and its derivatives. The other guanidinylating [326] (or guanylating [321,327]) reagents are cyanamides, carbodiimides, etc. [321,327,328]. Among them, thiourea oxides play an important role.

In 1955, Walter found that reaction of thiourea dioxide with glycine in concentrated ammonium hydroxide gave a 36% yield of guanidinoacetic acid [329]. Alanine, 4-aminobutanoic acid and norleucine also reacted to give the corresponding guanidine compounds, but no yields were reported. Miller and coworkers found that when the base was changed to 1 M potassium carbonate, no guanidinoacetic acid was produced [39]. The reaction of TDO with 1 M potassium carbonate gave ca. 30% ammonia. Under the same conditions thiourea trioxide reacts with glycine to give 80% yield of guanidinoacetic acid and with potassium carbonate alone gave less than 5% titrable volatile base (ammonia). These data correlate with the composition of products from thiourea di- and trioxides under basic conditions: dioxide produces mostly urea, but trioxide — cyanamide or carbodiimide, i.e. the more cyanamide is formed, the more guanidinoacetic acid can be synthesized in the reaction with glycine. Indeed, Walter reported [329] that control reaction of glycine with cyanamide under basic conditions yielded guanidinoacetic acid. It was also shown that glycine can be guanylated by N-phenylthiourea trioxide and its 2-methyl-, 2,6-dimethyl-, 2-chloro- and 2,6-dichloroderivatives, but rate of reaction with N-phenylthiourea trioxide is lower than with thiourea trioxide. This was attributed to a steric effect on the reactivity.

Interestingly, Miller and Bischoff found that reactions of amino acids with thiourea dioxides gave significantly poorer yields of guanidino acids than with thiourea trioxides (they studied guanylation reaction in presence of potassium carbonate) [92]. Nevertheless, some other authors used dioxides in the syntheses of guanidines. Thus, Knopp showed that compounds containing primary and secondary amino groups react with TDO to give the corresponding guanidines in alkaline solutions; for example, conversion of benzylamine to benzylguanidine proceeds almost quantitatively [330]. Jursic and coworkers have explored the reactions of amino acids with TDO in aqueous ammonia [331]. However, after mixing these reagents they received poor yields of guanidine derivatives, and separation of these products from the amino acids and inorganic salt formed in the course of reaction was also not an easy task. Changing the introduction order of the reagents (TDO was added to the solution of amino acid and sodium carbonate) and use of sodium carbonate can substantially increase the reaction conversion (up to 60%). Almost quantitative amino acid conversion was observed when sodium carbonate was replaced with sodium hydroxide. But with low-basic anilines such as 4-aminobenzoic acid, only 5–10% reaction conversion was observed. Therefore, authors believe that synthesis with TDO should not be practical for slightly nucleophilic amines, but should be the method of choice for conversion of aliphatic amines into the corresponding guanidine compounds. Thus, data of Miller [39] and Jursic [331] on the syntheses of guanidino compounds with TDO are contradictory.

Thiourea trioxides are more widely used in the preparations of guanidino compounds. The reaction of N-phenylthiourea trioxide with primary and secondary amines has been studied by Maryanoff and coworkers [93,94]. During the course of reaction of morpholine with N-phenylthiourea trioxide (phenylaminoiminomethanesulfonic acid) at room temperature, a transient intermediate was detected by TLC. When the reaction was studied by IR spectroscopy, no carbodiimide or cyanamide absorption was observed. Therefore, the authors suggested an addition/elimination mechanism involving the addition of the amine nucleophile to a thiourea trioxide to form a tetrahedral intermediate that collapses to product (Fig. 6.7):

Makarov and coworkers who studied the reaction of TDO with primary aliphatic amines (methylamine, isobutylamine) and ammonia came to the

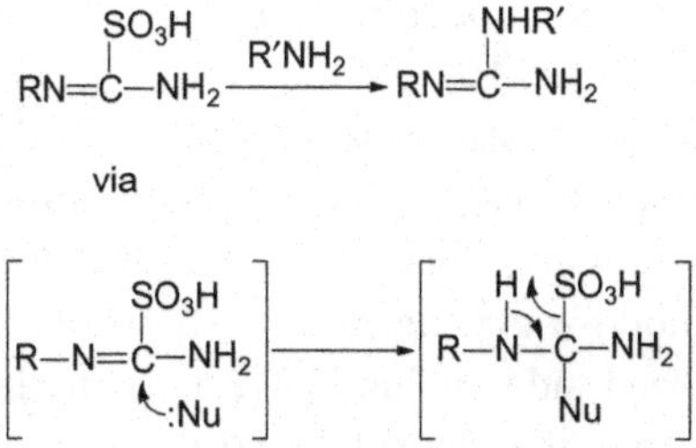

Fig. 6.7   Reaction of thiourea trioxide with amine

same conclusion [332]. The highest rate constants were observed for a reaction with methylamine, the lowest, for reaction with ammonia.

Monosubstituted guanidines have been synthesized with the yields of 50–80% from primary amines (cyclohexyl-, cyclooctyl-, n-butyl-, tert-butyl, phenyl-, 1,3-dimethylbutylamines, acosamine) and thiourea trioxide by Kim and coworkers (solvent — methanol) [96]. They found that reaction proceeds with greater facility than the displacement of alkyl mercaptan anion from S-alkylisothioureas in the classical guanidine synthetic procedure. In case of acosamine, reaction proceeded more rapidly and in a better yield (75%) in presence of three equivalents of triethylamine.

An interesting example of double guanidinylation and cyclization reaction of thiourea trioxide (aminoiminomethanesulfonic acid), (**1**) with methyl anthranilates (**2a**) to yield 5H-quinazolino [3,2-a] quinazoline-5,12 (6H)-diones (**3a**) was described by Prashad and coworkers [333]. Reaction of **2a** with **1** in methanol or ethanol did not lead to any guanidinylation, but treatment of **2a** with **1** in acetic acid yielded a mixture of acetate salts of **3a** and **4a** (Fig. 6.8). Similarly, a reaction of **1** with methyl 4-chloro anthranilate (**2b**) also furnished the corresponding 2,9-dichloro-5H-quinazolino[3,2-a]quinazoline-5,12 (6H)-dione (**3b**) and 7-chloro-2-amino-4(3H)-quinazolinone (**4b**).

N-hydroxyguanidines were synthesized by treating the corresponding thiourea trioxide with hydroxylamine and triethylamine [334]. Thus, N,N'-diphenylthiourea and N-ethyl-N-phenylthiourea trioxides gave a 37% yield of N''-hydroxy-N,N'-diphenylguanidine (**5a**) and 24% yield of N''-hydroxy-N-ethyl-N-phenylguanidine (**5b**), respectively. Treatment of N-phenylthiourea, N,N'-diethylthiourea and N-ethyl-N-phenylthiourea trioxides with cyanamide led to 30–55% yields of the corresponding N-cyanoguanidines (**6**) (see Fig. 6.9).

Thiourea oxides are widely applied reagents in the syntheses of bioactive guanidines. Thus, thiourea trioxide was used in the synthesis of thrombin inhibitor DuP 714 (see Fig. 6.10) [335].

The attempts to synthesize this compound using other guanylating compounds (O-methylisourea, S-methylisothiourea and 3,5-dimethyl-1-formamidinopyrazole) failed. One year later, Kettner and his coworkers developed a new method for asymmetric synthesis of α-aminoboronic acids containing functionalized side chains [336]. This method included guanylation reaction with thiourea trioxide according to previously published procedure [335]. In 1999, Kettner's group had synthesized the other

(1)

(2a) R = H
(2b) R = Cl

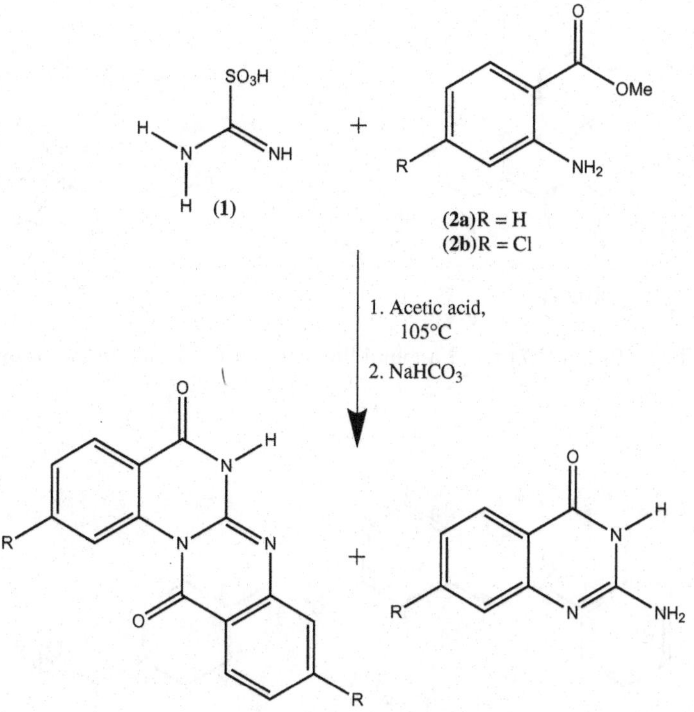

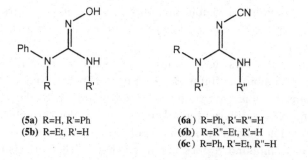

(3a) R=H
(3b) R=Cl

(4a) R=H
(4b) R=Cl

Fig. 6.8 Reaction of thiourea trioxide with methyl anthranilates

(5a) R=H, R'=Ph
(5b) R=Et, R'=H

(6a) R=Ph, R'=R''=H
(6b) R=R''=Et, R'=H
(6c) R=Ph, R'=Et, R''=H

Fig. 6.9 Guanidines synthesized from thiourea trioxide

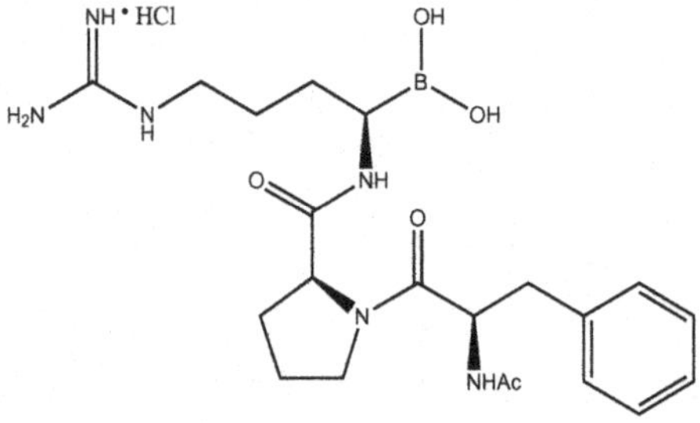

Fig. 6.10   DuP 714 and a guanidino-containing thrombin inhibitor

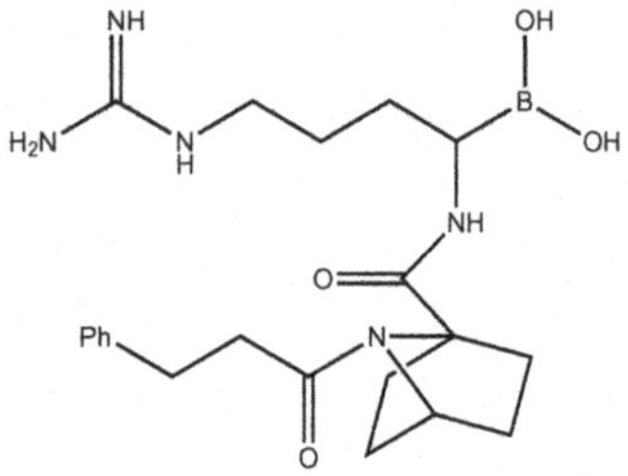

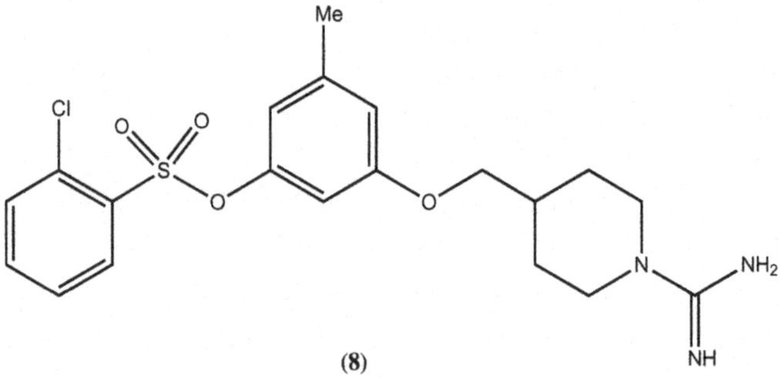

Fig. 6.11   A guanidino-containing thrombin inhibitor synthesized by Lu *et al.*

guanidino-containing thrombin inhibitor **7**, which incorporates conformationally constrained 7-azabicycloheptane carboxylic acid as a proline displacement [337].

Lu and coworkers have synthesized another effective guanidino-containing thrombin inhibitor **8** (see Fig. 6.11) [338]. For conversion of amino group to guanidine group they used thiourea trioxide.

This inhibitor exhibited remarkable specificity for thrombin. The compound showed no detectable inhibition of plasmin, chymotrypsin, urokinase or elastase. Compound **8** does not interact with the active-site catalytic apparatus and is anchored to the enzyme via a single network of hydrogen bond to Asp189 of the S1 pocket. Authors showed that rigid guanidine backbone is optimal for activity [338,339]. Thiourea trioxide has also been used in the synthesis of thrombin inhibitors by Iwanowicz and coworkers [340].

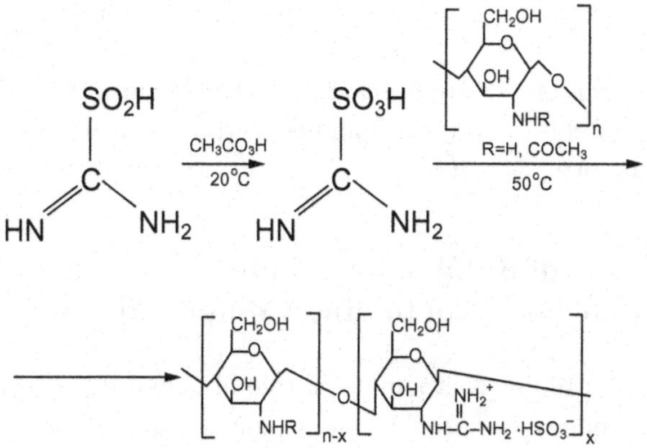

Fig. 6.12   Reaction of thiourea trioxide with chitosan

Colanduoni and Villafranca showed that thiourea dioxide when incubated with *E.coli* glutamine synthetase leads to inactivation by irreversible covalent modification [341]. Homoarginine is produced as a result of this reaction, as determined by amino acid analysis. Thiourea trioxide reacted with the same lysine residues of glutamine synthetase as thiourea dioxide [342]. Thiourea dioxide was also used in chemical modification studies to identify functionally important amino acids in *E.coli* cytidine triphosphate (CTP) synthetase [343]. Incubation at pH = 8.0 in the absence of substrates led to rapid, time dependent and irreversible inactivation of the enzyme. The specificity of TDO for lysine residues indicates that one or more lysines are most likely involved in CTP synthetase activity.

Thiourea oxides can also be used for modification of polymers containing amino groups. The most important example of such a polymer is chitosan. It is composed of glucosamine units with free amino group on the second carbon, which can be used as the reactive site [344]. Guanidinylated chitosan derivatives with different molecular weights have been synthesized by the guanidinylation reaction of chitosan with thiourea trioxide (Fig. 6.12) [345]. It was shown that compared to chitosan, guanidinylated chitosan had much better antibacterial activity. The antibacterial activity of guanidinylated chitosan enhances with the decreasing pH [345].

Zhai and coworkers showed that guanidinylated chitosan is a promising candidate as an effective non-viral vector for *in vivo* gene delivery [346]. One year later, guanidinylated chitosan was tested for siRNA delivery [347]. Authors have demonstrated a potential application of the chitosan-derived nanodelivery vehicle in RNA interference therapy for lung diseases via aerosol inhalation.

The other examples of application of thiourea oxides in the syntheses of bioactive compounds containing guanidino groups can be found in several independent reports [348–360].

6.3   Synthesis of Sulfur-, Selenium- and Tellurium-Containing Compounds

Being sulfur-containing compounds, dithionite, hydroxymethanesulfinate and thiourea oxides can "provide" sulfur-oxygen fragments to other compounds. This property is used in the synthesis of sulfones. Over the last 20 years, usage of sulfones in organic synthesis has increased dramatically, the synthetic opportunities of sulfones have been developed to such an extent as to rival the carbonyl functionality for versatility [361].

Sodium hydroxymethanesulfinate (rongalite) was first used in the synthesis of dibenzyl sulfone from benzyl chloride in alcoholic sodium hydroxide in 1908 [362]. In 1964, Wellisch and coworkers [363] prepared several sulfones and polysulfones by the reaction of sodium dithionite with alkyl halides and dihalides in DMF at 110°C for 9 h in a yield of 17% (for reaction with RCl), according to the following equation:

$$2RX + Na_2S_2O_4 \longrightarrow RSO_2R + SO_2 + 2NaX. \qquad (6.14)$$

However, this procedure has suffered from disadvantages such as limited number of substrates, low yield and long reaction time.

Amiri and Mellor have used DMSO as a solvent for reaction of α-chloromethylnaphthalene with sodium dithionite at 110°C for 9 h, but the product (di-α-methylenenaphtylsulfone) has also been received in a low yield (18%) [364]. Kerber and Starnick [365] have received β-β'-disubstituted diethyl sulfones by reaction of vinyl compounds in which the olefinic bond is polarized by carbonyl, pyridyl, nitrile or sulfonyl groups with hydroxymethanesulfinate, thiourea dioxide and dithionite. For example, reaction with hydroxymethanesulfinate can be described in Fig. 6.13 [365].

Synthesis from dithionite and TDO also proceeds with intermediate formation of sulfoxylate.

The reaction of hydroxymethanesulfinate with 1,4- and 1,2-benzoquinone as well as with 1,4-naphthoquinone, which can be generated *in situ* from the corresponding dihydroxyaryl compounds by oxidation, yields

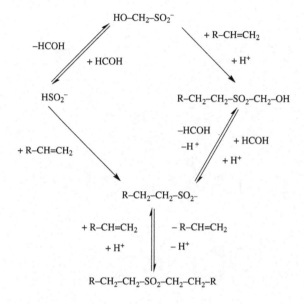

Fig. 6.13   Synthesis of β-β'-disubstituted diethyl sulfones

Fig. 6.14   Reaction of 1,4-benzoquinone with hydroxymethanesulfinate

symmetrical bis(dihydroxyaryl) sulfones [366]. For example, reaction with 1,4-benzoquinone can be described in Fig. 6.14.

Messinger and Greve [40] have synthesized with yields of 14–66% symmetrical sulfones $R-CH_2-SO_2-CH_2-R$ from sodium hydroxymethanesulfinate and Mannich bases $R-CH_2-N(CH_3)_2$ (R is shown in Fig. 6.15).

The same authors synthesized the cyclic sulfone in 14% yield when 4-(dimethylamino)-1,3-diphenylbutan-2-one hydrochloride was heated with sodium hydroxymethanesulfinate in aqueous medium (Fig. 6.16) [367].

In 1988, Harris has suggested a simple, one step procedure for the preparation of dibenzyl sulfones from benzyl halides using sodium

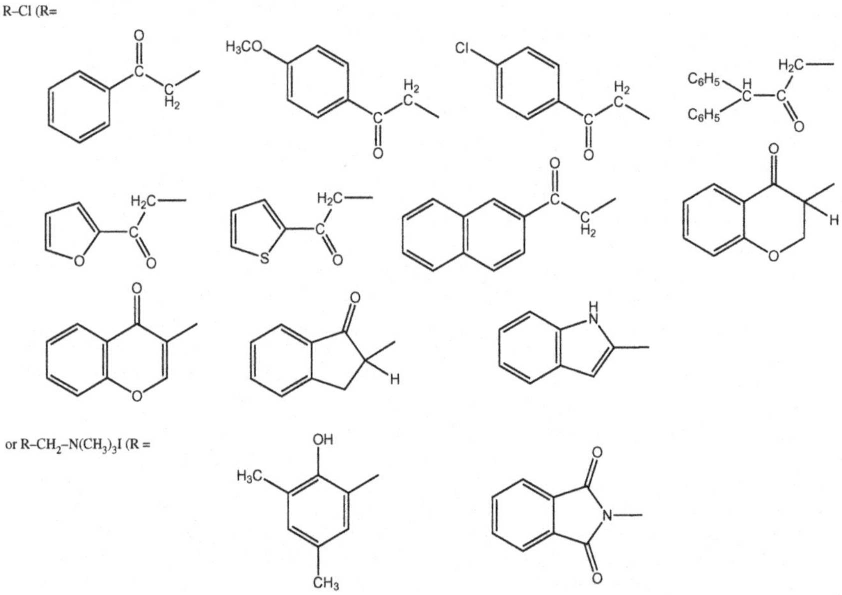

R–Cl (R=

or R–CH$_2$–N(CH$_3$)$_3$I (R =

Fig. 6.15   Synthesis of symmetrical sulfones R$-$CH$_2$$-SO_2$$-CH_2$$-$R from sodium hydroxymethanesulfinate and Mannich bases

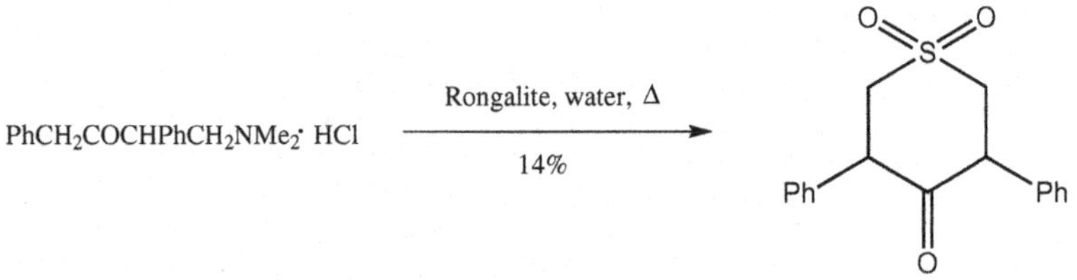

PhCH$_2$COCHPhCH$_2$NMe$_2$ · HCl

$\xrightarrow[14\%]{\text{Rongalite, water, } \Delta}$

Fig. 6.16   Synthesis of cyclic sulfone

hydroxymethanesulfinate and dimethylformamide as a solvent (yields 35–48%) [368]. The reaction presumably proceeds via the benzyl sulfinate (Fig. 6.17).

Intermediate sulfinate can be formed in the reaction between benzyl halide with sulfoxylate or hydroxymethanesulfinate, followed by loss of formaldehyde. Attempts were made to monitor the reaction by NMR, but no intermediates could be obtained.

One year later, Harris and coworkers have improved method of synthesis of dibenzyl sulfones [369]. They have suggested to prepare these

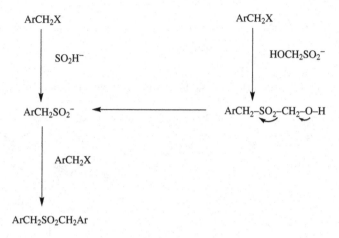

Fig. 6.17   Synthesis of dibenzyl sulfones

compounds from the corresponding benzyl halide under mild conditions using either sodium dithionite/Aliquat (phase transfer reagent), or sodium hydroxymethanesulfinate/potassium carbonate in the absence of solvent. Both of these procedures give higher yields under milder conditions, and are much simpler to be carried out than the previously reported methods. Thus, dibenzyl sulfone has been received from benzyl bromide in yields 61% and 76% using dithionite and hydroxymethanesulfinate, respectively.

Primary benzyl and allyl halides (bromides and/or chlorides) in reaction with sodium dithionite in DMF or HMPA gave the reductively coupled dimers and/or sulfones [370]:

$$2\,ArCH_2X \xrightarrow{155-160°C} ArCH_2CH_2Ar \text{ and/or } ArCH_2SO_2CH_2Ar. \quad (6.15)$$

Reductive coupling of secondary benzylic halides (bromides and chlorides) gave the corresponding reductively coupled dimers:

$$2\,Ar_2CHX \xrightarrow{155°C} Ar_2CHCHAr_2. \quad (6.16)$$

Hoey and Dittmer have received symmetrical sulfones by the treatment of benzyl bromides with rongalite at 80°C in aqueous DMF in the presence of potassium bicarbonate (a yield 45–88%) [371]. Also, α,α'-dibromo-o-xylene was found to react with rongalite–SO_2 in DMF at 70°C to furnish sulfone in 75% yield (Fig. 6.18) [371].

Kotha and coworkers have demonstrated the synthesis of highly functionalized benzosulfones via rongalite [372]. The dibromide was treated with

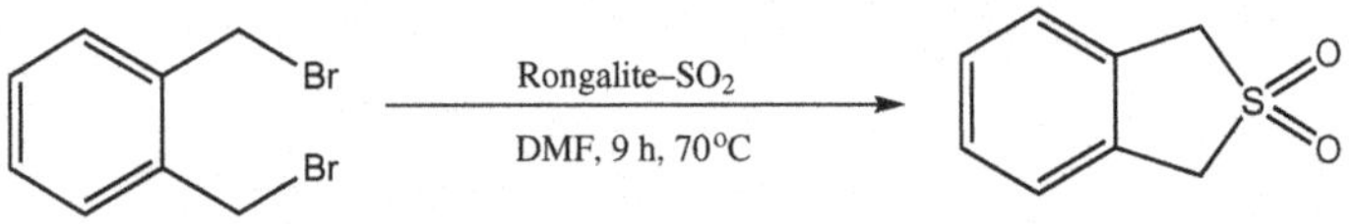

Fig. 6.18   Synthesis of sulfone from α, α'-dibromo-o-xylene

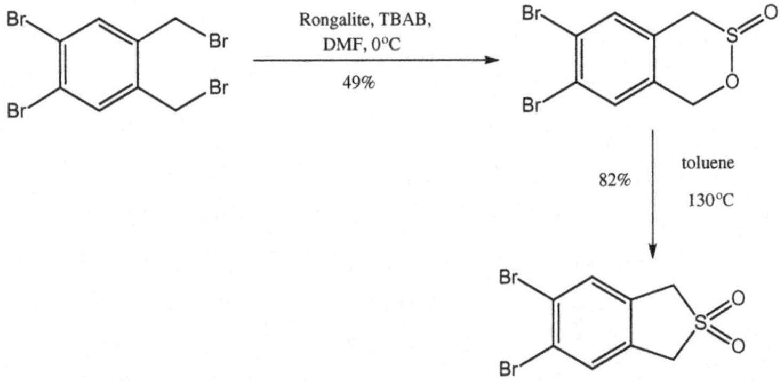

Fig. 6.19   Synthesis of benzosulfones via rongalite in the presence of TBAB

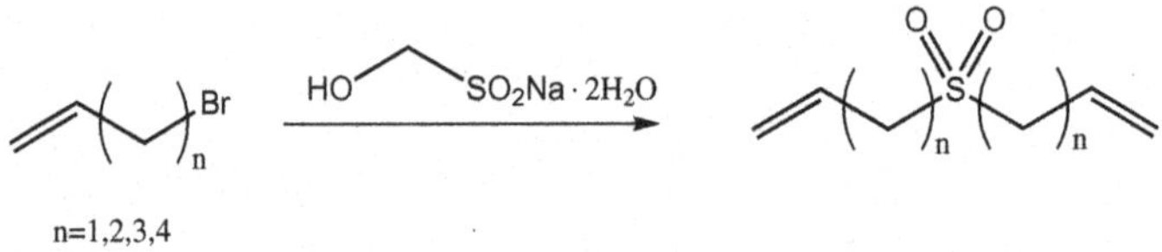

Fig. 6.20   Synthesis of olefinic sulfones from alkenyl bromides and rongalite

rongalite in the presence of tetrabutylammonium bromide (TBAB) as a phase-transfer catalyst in DMF to deliver the sultine derivative in 49% yield. Its rearrangement under thermal conditions gave the corresponding sulfone (Fig. 6.19).

In another example [373], the open-chain, terminally olefinic sulfones have been obtained by treatment of alkenyl bromides with rongalite. The reaction has been performed at room temperature in DMF, and potassium carbonate has been used as a base in the presence of TBAB (Fig. 6.20).

The possible mechanism for the formation of sulfone derivatives is shown in Fig. 6.21. The reaction involves nucleophilic displacement by the hydroxymethanesulfinate anion, followed by the loss of a formaldehyde molecule

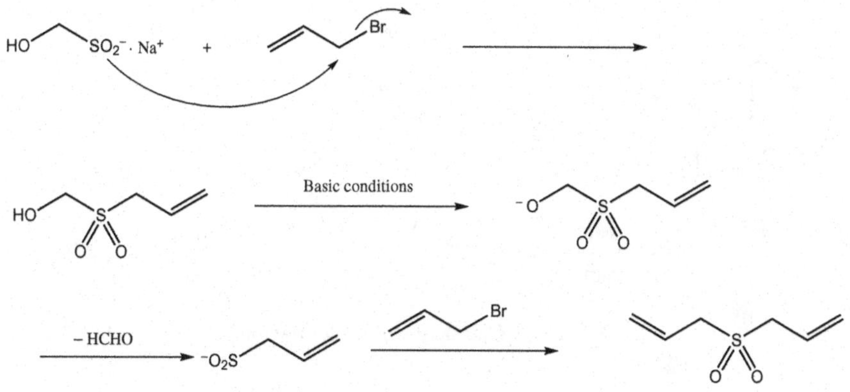

Fig. 6.21 Proposed mechanism for reaction of alkenyl bromides and rongalite

in the presence of a base, thus generating a nucleophile for the second alkylation step.

Li and Zhang have synthesized symmetric dibenzyl sulfones in ionic liquid 1-butyl-3-methylimidazolium tetrafluoroborate ([bmim]BF$_4$) at 100°C from various benzyl chlorides and sodium dithionite in moderate yields (50–78%) [374]. In most cases, the reaction was completed in about 4–6 h at 100°C. The advantage of this method compared with the previous results in DMF [363] and DMSO [364], besides yields and rates, is that the ionic liquid is recyclable.

Sodium dithionite is used not only in the synthesis of sulfones, but also in their transformations to other compounds [375,376]. The most important and well-known transformation is the synthesis of olefins from sulfones. In 1973, Marc Julia and Jean-Marc Paris reported a novel olefin synthesis which utilized the reductive elimination of β-acyloxysulfones as an alkene forming step [377]. Alkene formation via the classical Julia reaction is a relatively cumbersome affair and typically requires four distinct synthetic operations (Fig. 6.22): metallation of a phenylsulfone (Step 1), addition of the metallate to an aldehyde (Step 2), acylation of the resulting β-alkoxysulfone (Step 3), and reductive elimination of the β-acyloxysulfone with a single electron donor to afford alkene products (Step 4) [378].

In 1982, Julia and coworkers have shown that E or Z vinylsulfones can be reduced stereospecifically to the corresponding olefins with sodium dithionite (Fig. 6.23) [375].

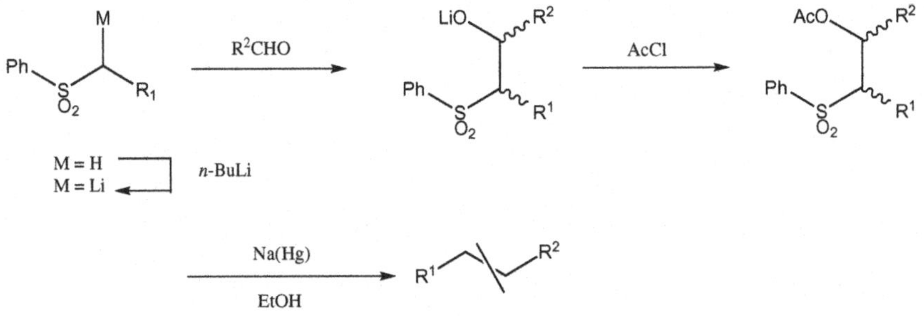

Fig. 6.22   Classical Julia olefination

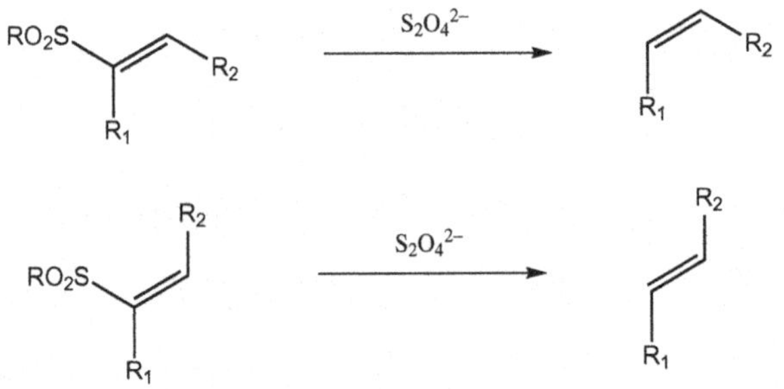

Fig. 6.23   Synthesis of olefins from vinylsulfones

Later they studied the mechanism of this reaction [376]. To discriminate between electron transfer and addition-desulfonylation-elimination mechanisms Julia and coworkers tried a number of reducing conditions which are known to transfer electrons to substrate: sodium amalgam in methanol, electrolysis, sodium or lithium in liquid ammonia, lithium in ethylamine. In all cases mixtures of E and Z olefins were formed from both isomers of the vinylic sulfone. Vinyl radicals are intermediates, and these are known to equilibrate very rapidly, apparently faster than the second electron is transferred. The vinyl anions are configurationally stable. This made it unlikely that the stereospecific hydrogenolysis with sodium dithionite could take place via successive electron transfers since it was difficult to believe that this rather weak reducing agent could possibly reduce a vinyl radical to the corresponding anion faster than reductants mentioned earlier. A "nucleophilic" reduction process was therefore considered. Authors assumed that the species $H^+ + SO_2^{2-}$ or HSO_2^- could behave as a hydride ion, in other

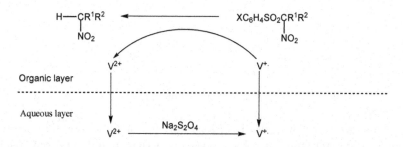

Fig. 6.24 Cyclic pathway for the viologen-mediated reductive desulfonylation of α-nitro-sulfones by sodium dithionite (V^{2+}–viologens (1,1'-dialkyl-4,4'-bipyridiniums))

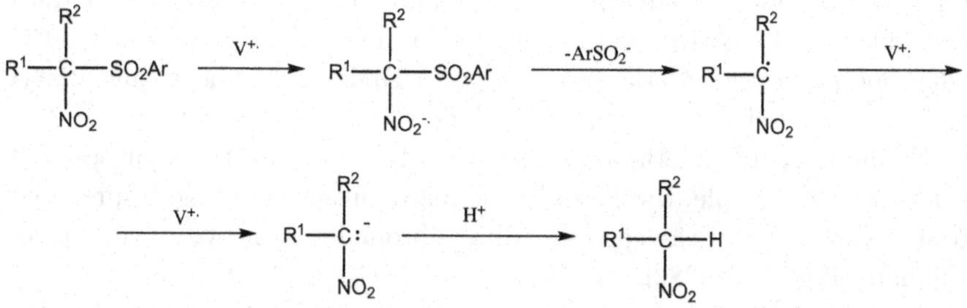

$$R^1\!-\!\underset{\underset{NO_2}{|}}{\overset{\overset{R^2}{|}}{C}}\!-\!SO_2Ar \xrightarrow{V^{+\cdot}} R^1\!-\!\underset{\underset{NO_2^{-\cdot}}{|}}{\overset{\overset{R^2}{|}}{C}}\!-\!SO_2Ar \xrightarrow{-ArSO_2^{-}} R^1\!-\!\underset{\underset{NO_2}{|}}{\overset{\overset{R^2}{|}}{\overset{\cdot}{C}}} \xrightarrow{V^{+\cdot}}$$

$$\xrightarrow{V^{+\cdot}} R^1\!-\!\underset{\underset{NO_2}{|}}{\overset{\overset{R^2}{|}}{C}}\!:^{-} \xrightarrow{H^+} R^1\!-\!\underset{\underset{NO_2}{|}}{\overset{\overset{R^2}{|}}{C}}\!-\!H$$

Fig. 6.25 Proposed mechanism for the viologen-mediated desulfonylation of α-nitro sulfones

words, sulfoxylate would be the two electron carrier necessary for converting H^+ into H^-. The overall reduction process is equivalent to a nucleophilic substitution process by a hydride ion on a vinylic substrate. This can proceed in two ways: α-addition-elimination or β-addition-elimination. Julia and coworkers found that hydrogenolysis proceeds via intermediate which could be isolated after alkylation *in situ* to 1,2-bissulfone. This result showed that the mechanism is β-addition-elimination type.

Park and coworkers studied viologen-mediated reductive desulfonylation of α-nitro sulfones to corresponding nitro compounds by sodium dithionite in organic solvent-water two-phase systems (Fig. 6.24) [379].

The sulfones which do not have an α-nitro group, e.g. benzyl phenyl sulfone, could not be desulfonylated under these experimental conditions. The proposed mechanism of desulfonylation is shown in Fig. 6.25. The reduction is initiated by a SET from $V^{+\bullet}$ to α-nitro sulfones.

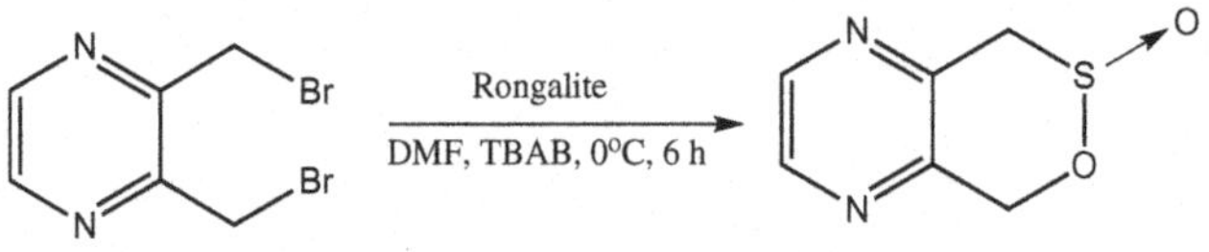

Fig. 6.26   Synthesis of sultines

Thiourea dioxide was also used for reduction of sulfur-containing compounds. Borgogno and coworkers have found that sulfilimines and disulfides are reduced by TDO to the corresponding sulfides and thiols under two-phase conditions using hexadecyltributylphosphonium bromide as a phase-transfer catalyst. Their attempts to reduce sulfoxides (methyl p-tolyl sulfoxide and diphenyl sulfoxide) under these conditions were unsuccessful [380]. Drabowicz and Mikolajczyk found that it is possible to reduce sulfoxides to the corresponding sulfides by TDO in acetonitrile at 80°C when iodine is used as the catalyst [381]. Dialkyl, alkyl-aryl, and diaryl sulfoxides are reduced in yields exceeding 90%.

Sodium hydroxymethanesulfinate is also used in the synthesis of sultines. For example, the heterocyclic-fused sultines are readily prepared from the reaction of the corresponding dibromides with hydroxymethanesulfinate (Fig. 6.26) [382].

In organotellurium chemistry, sodium hydroxymethanesulfinate has the widest application. This application has a long history which began from the paper of Tschugaeff and Chlopin. They found that on heating with an excess of sodium hydroxymethanesulfinate in dilute aqueous sodium hydroxide under nitrogen, tellurium is reduced to give sodium telluride Na_2Te_2 [383]. Using their procedure, Balfe and coworkers have synthesized di-n-butyl telluride from tellurium, hydroxymethanesulfinate and n-butyl bromide [384]. McCullough has received (see Fig. 6.27)1,4-thiatellurane from the reaction between β, β'-dichlorodiethyl sulfide (mustard gas) and Te−HOCH$_2$SO$_2$Na solution [385]. Tellurophene was synthesized by using bis(trimethy1-silyl)-1,3-butadiyne and sodium telluride generated *in situ* from tellurium and sodium hydroxymethanesulfinate [386].

Vicinal dibromoalkanes are debrominated to alkenes by treatment with Te−HOCH$_2$SO$_2$Na solution [387]:

$$RCHBr\text{-}CHBrR' \xrightarrow{Na_2Te, room\ temp.} RCH = CHR'. \qquad (6.17)$$

The reaction is of considerable synthetic interest, particularly when a molecule contains other groups such as carbonyl, carboxyl, ester and nitro groups because all of these remain unaffected during the debromination.

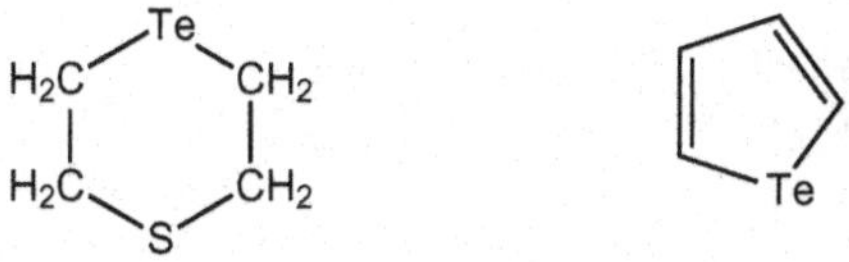

1,4 thiatellurane   tellurophene

Fig. 6.27   Structure of 1,4-thiatellurane and tellurophene

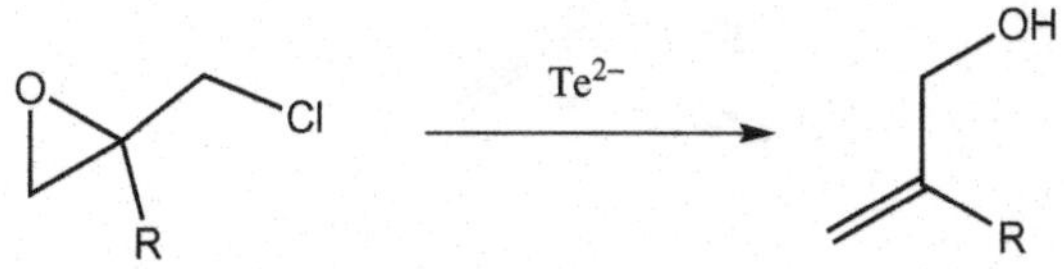

Fig. 6.28   Selective formation of allyl alcohols

Sodium telluride easily reduces aromatic nitro compounds to the corresponding amines in good yields. The reduction can be carried out by using catalytic amount of tellurium, since sodium telluride is readily regenerated in presence of excess rongalite [388].

A series of works dedicated to application of rongalite in tellurium chemistry has been done by Dittmer and coworkers [389–395]. In studies of reactions of sulfide, selenide and telluride ions with chloromethyloxiranes, they have observed that whereas the sulfide dianion preferentially forms the four-membered thietanes, selenide ion forms both the four-membered heterocycle and a non-selenium containing allyl alcohol. Telluride ion (received from $Te-HOCH_2SO_2Na$ reaction in alkaline aqueous solution), in contrast, forms only the allyl alcohol (see Fig. 6.28), without any trace of the four-membered ring [389].

Under appropriate reaction conditions, acetylenic substituted chloromethyloxiranes are converted in good yields (ca. 59–73%) to the corresponding tellurophenes, the allyl alcohol being produced as by-product (see Fig. 6.29) in the reaction [390].

It was found that nature of telluride ion in solution, which depends on the type of reductant using for its preparation, has a dramatic effect on the ratio of the two products (tellurophen and alcohol) formed. If sodium borohydride is used, monomeric telluride ion, Te^{2-} or $[(MeO)_3BTe]^{2-}$ in equilibrium with Te^{2-} are the main products. Reduction of Te with sodium hydroxymethanesulfinate leads to predominant formation of a variety of

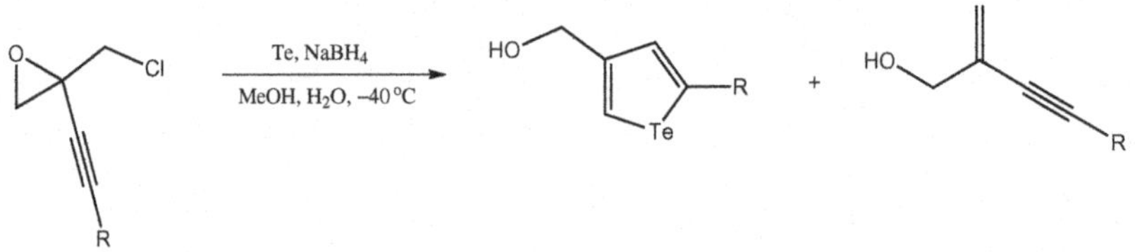

Fig. 6.29   Parallel formation of tellurophenes and allyl alcohols

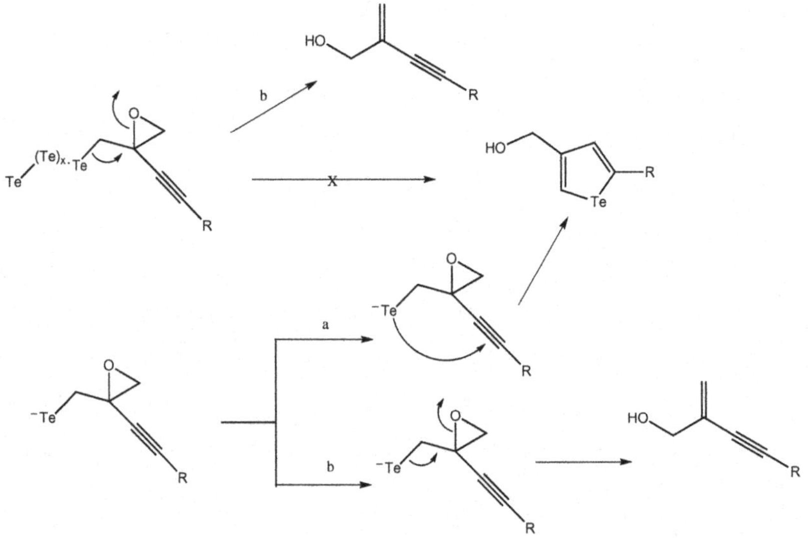

Fig. 6.30   Synthesis of tellurophene and allylic alcohol

polytellurides, $-Te(Te)_x Te^-$. Therefore, in the first case acetylenic substituted chloromethyloxiranes produce predominantly tellurophene, but in the
second case, allylic alcohol (Fig. 6.30) [390].

Irgolic and his coworkers have prepared tetrahydroselenophene and
tetrahydrotellurophene in yields between 80% and 90% by the reaction
between disodium chalcogenides and 1,4-dibromobutane in a medium
of aqueous sodium hydroxide in the presence of methyltrialkyl (C_8–
C_{10}) ammonium chloride as the phase-transfer catalyst [391]. The disodium chalcogenides synthesized from the elemental chalcogens and sodium
hydroxymethanesulfinate in aqueous sodium hydroxide solution were used
in situ (Fig. 6.31).

Dittmer and coworkers have used tellurium–rongalite system in the synthesis of diverse allylic alcohols [392–396]. Tellurium is reduced to telluride

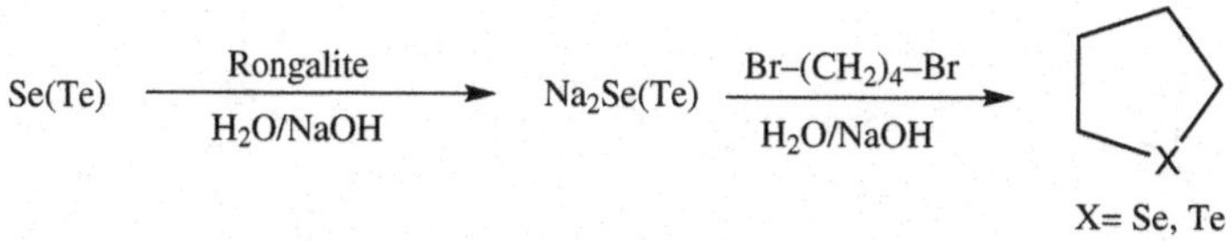

Fig. 6.31   Synthesis of tetrahydroselenophene and tetrahydrotellurophene

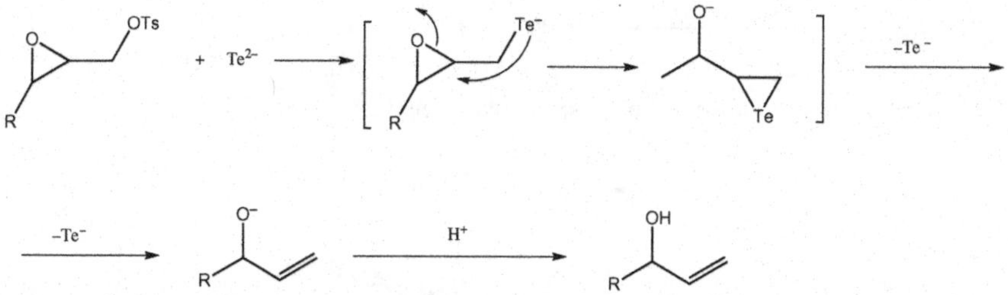

Fig. 6.32   Proposed mechanism of the synthesis of allylic alcohols

ions Te_n^{2-} (n = 1, 2, 3), which triggers the hypothetical reaction sequence shown in Fig. 6.32. It then results in the formation of the desired product with the release of tellurium which can be readily reused. The process may be performed under no-solvent or phase transfer conditions [395] and can be catalytic in tellurium [394]. It was also shown that telluride ion reacts with an oxiranemethanol toluenesulfonate in toluene under phase-transfer conditions to give an allylic alkoxide anion that is trapped by addition to an adjacent α, β-unsaturated ester to yield a furan derivative [396]. Use of a non-racemic oxiranemethanol tosylate gives two furan diastereomers in a ratio of 56:44. Tellurium-promoted reductive-epoxide ring-opening cascade reaction has been utilized also in the synthesis of the marine sponge diterpenoid nakamurol A [397].

Rongalite is widely used in the syntheses of sulfur- and seleno-containing compounds. Series of papers were dedicated to reactions of disulfides and diselenides. In 2009, Guo and coworkers reported a highly efficient and regioselective synthesis of β-hydroxy sulfides by rongalite promoted thiolysis of epoxides with disulfides (Fig. 6.33) [398].

A tentative mechanism of the formation of β-hydroxy sulfides is shown in Fig. 6.34.

A year later, the same group developed the synthesis of β-sulfido carbonyl compounds through rongalite and base-promoted cleavage of disulfide

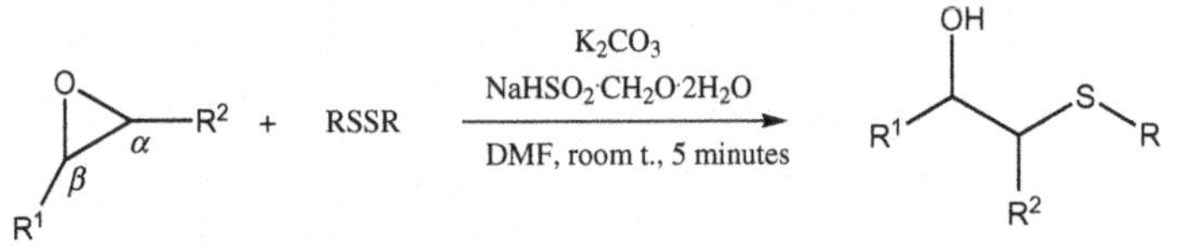

Fig. 6.33   Synthesis of β-hydroxy sulfides

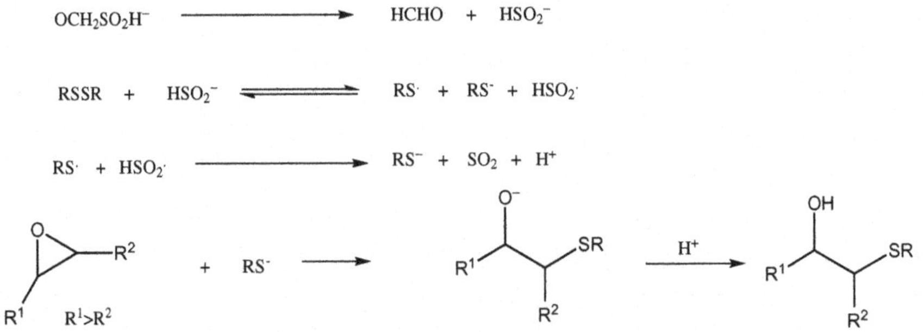

Fig. 6.34   Proposed mechanism of synthesis of β-hydroxy sulfides

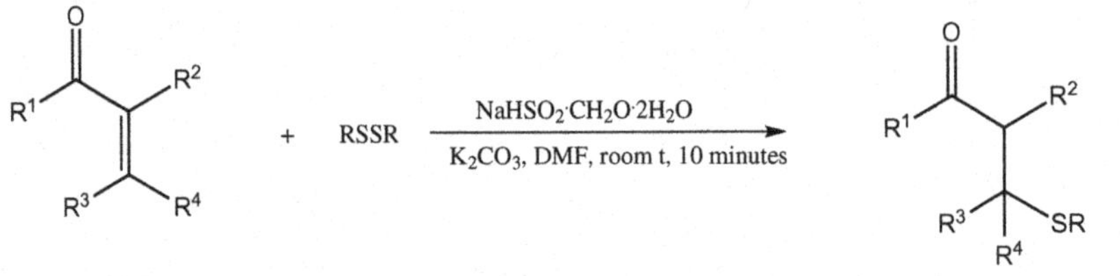

Fig. 6.35   Synthesis of β-sulfido carbonyl compounds

and subsequent Michael addition to α,β-unsaturated ketones/esters (Fig. 6.35) [399]. As in the synthesis of β-hydroxy sulfides, the intermediate of the process is sulfoxylate ion.

Rongalite was used in one-pot synthesis of β-amino/β-hydroxy selenides and sulfides from aziridines and epoxides [400]. In this method, diaryl disulfides and diselenides undergo facile cleavage on treatment with rongalite to generate the corresponding thiolate and selenolate species *in situ*, which affect the ring openings of aziridines and epoxides in a regioselective manner. Wang and coworkers have developed an efficient and stereoselective

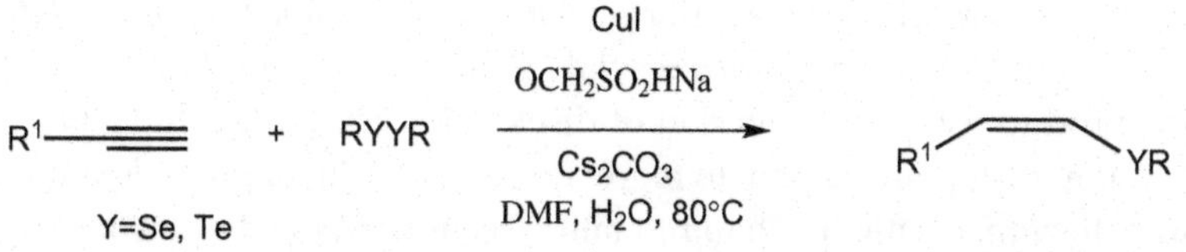

Fig. 6.36 Stereoselective protocol for hydrothiolation of terminal alkynes

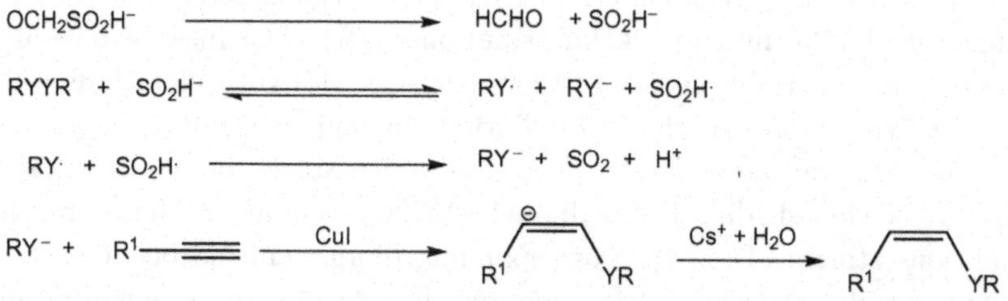

Fig. 6.37 Possible mechanism of synthesis of (Z)-1-alkenyl sulfides (selenides)

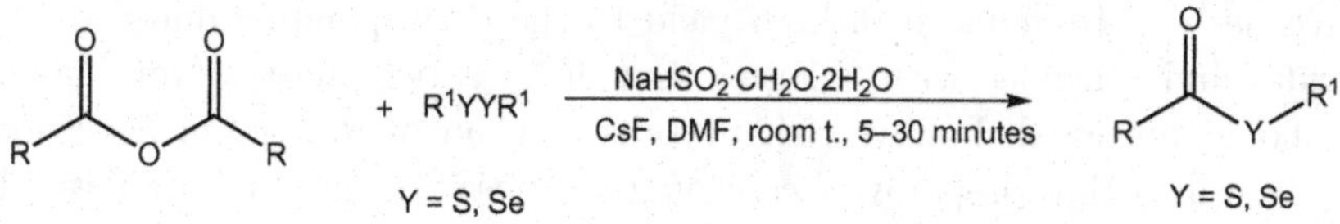

Fig. 6.38 Synthesis of thioesters and selenoesters

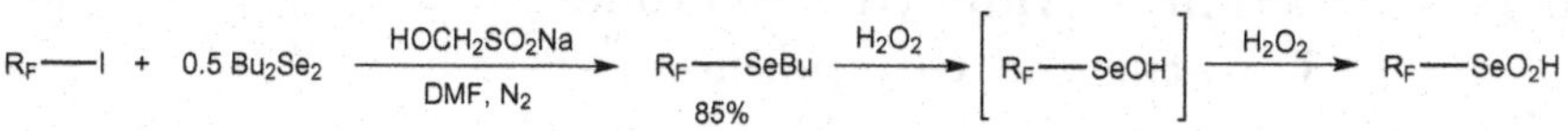

Fig. 6.39 Synthesis of fluorous seleninic acid

protocol for the hydrothiolation of terminal alkynes with diaryl disulfides and 1,2-diphenyl diselenide (see Fig. 6.36) [401].

A possible mechanism of reaction is shown in Fig. 6.37.

Dan and coworkers have developed one-pot synthesis of thioesters and selenoesters promoted by rongalite (see Fig. 6.38) [402].

Rongalite was also used in the syntheses of fluorous seleninic acid (Fig. 6.39) [403] and selena-fatty acids [404].

Reich and coworkers have reported the synthesis of alkyl-, aryl selenides from the appropriate halide or mesylate using $ArSe^-$ in ethanol, which in turn is produced by the reduction of diselenides using rongalite [405].

Sodium dithionite is much more rarely used in organoselenium and organotellurium chemistry than rongalite, seemingly, due to instability, sensitivity to dioxygen and insolubility in non-aqueous solutions. Note, however, that selenium itself can be produced from the reaction between sodium dithionite and selenious acid in acidic aqueous solution [406].

Similar to hydroxymethanesulfinate, thiourea dioxide can be used for reduction of tellurium in alkaline solutions [407]. The intermediate disodium telluride or ditelluride further reacts *in situ* with alkyl halides to give dialkyltellurides or dialkylditellurides in high yield. Comasseto and coworkers have also described a procedure which allows for the *in situ* synthesis of arylalkyl, diaryl and dialkylchalcogenides under phase transfer conditions starting from the corresponding diorganodichalcogenides. The dichalcogenides are reduced by thiourea dioxide in alkaline medium and catalyzed by a quaternary ammonium salt [408]. A year later, Lang and Comasseto showed that aryltellurium trihalides, arylselenium trihalides, organyltellurium and selenium dichlorides and organylselenoxides and telluroxides can be reduced in high yield to the corresponding diaryl ditellurides and selenides with thiourea dioxide in a two-phase system [409]. It should be noted, however, that works of Comasseto's group have not received a further preparative continuation, possibly, due to instability of thiourea dioxide in alkaline solutions.

6.4   Reduction of Aldehydes, Ketones and Unsaturated Compounds

Reaction of sodium dithionite with aldehydes occupies a special place in the chemistry of sulfur-containing reductants, since on the one hand, this is a method of synthesis of α-hydroxyalkane(arene)sulfinates, including hydroxymethanesulfinate (rongalite), but on the other hand, method of synthesis of alcohols [15]. In the second case, corresponding sulfinates are the intermediates of the reaction.

Mulliez and Naudy have synthesized α-hydroxyalkane(arene)sulfinates by slowly adding aldehydes to the mixture sodium dithionite–sodium hydroxide under argon (Eq. (6.18)) [71]. Since corresponding sulfonate is unstable in alkaline solutions, it does not form from dithionite. Sulfinates are rapidly separated from sodium sulfite and isolated. Separation is based on the different solubility of sulfinates and sulfite in water (in the case of aromatic derivatives) and in ethanol (in the case of aliphatic derivatives):

$$S_2O_4^{2-} + RCOH + OH^- \longrightarrow HOCHRSO_2^- + SO_3^{2-}, \qquad (6.18)$$

where R = Ph, Me, p-C_6H_4, p-$CF_3C_6H_4$, CF_3, and the corresponding yields are 86%, 78%, 40%, 80%, 80%, respectively.

In most of the other studies reaction of dithionite with aldehydes was used for the synthesis of alcohols. De Vries and coworkers have shown that with excess dithionite in H_2O/dioxane simple aldehydes and ketones can be reduced at reflux temperature (Eq. (6.19)) [51,410]. Their data have been confirmed by Minato and coworkers [411] studying the reaction in refluxing MeOH–H_2O.

$$R^1C(O)R^2 + S_2O_4^{2-} \xrightarrow{2H_2O} R^1CH(OH)R^2 + 2HSO_3^-, \qquad (6.19)$$

where R^1 = alkyl, aryl and R^2 = alkyl, aryl or H, respectively.

Some aliphatic ketones, for example, pentanone and 4-heptanone, are reduced in better yields by adding dimethylformamide to the reaction mixture, again held at reflux [51]. α-Hydroxyalkane(arene)sulfinates are suggested as probable intermediates in these reductions (Eq. (6.20)).

$$R^1C(O)R^2 + S_2O_4^{2-} \underset{}{\overset{H_2O/\text{base}}{\rightleftharpoons}} R^1C(OH)(SO_2^-)R^2 + HSO_3^-. \qquad (6.20)$$

For examination of the possibility that α-hydroxyalkane(arene)sulfinates could be intermediates in the reductions reported here, derivative of benzaldehyde was synthesized. This was added to a refluxing mixture of dioxane–H_2O. The reaction product consisted of a 65/35 mixture of benzyl alcohol and benzaldehyde. The formation of sulfinate derivative of benzaldehyde is known to be reversible explaining the presence of benzaldehyde (Eq. (6.21)). The results provide an evidence for intermediate formation of sulfinate in the reduction of benzaldehyde. Data of de Vries and coworkers also show that sulfinate derivative should be sufficiently unstable to allow reductive decomposition with loss of SO_2 (HSO_3^-). Indeed, they have received alcohols from such aldehydes as n-hexanal, benzaldehyde, 2-furyl aldehyde as well as ketones (2-octanone, etc.) which form relatively unstable sulfonate derivatives. This conclusion can be done on the basis of comparative stability of corresponding sulfinates (see, for example, Ref. [412], where authors have shown that benzaldehyde–S(IV) adduct is less stable than formaldehyde–S(IV) adduct). It should be noted, however, that authors left open the question which species — sulfoxylate or sulfur dioxide anion radical — reacts with aldehyde,

$$C_6H_5CHO + S_2O_4^{2-} \xrightleftharpoons{\text{r.t.}} C_6H_5CH(OH)SO_2^-$$
$$\xrightarrow{\text{dioxane/H}_2\text{O}} C_6H_5CH_2OH + HSO_3^-.$$

(6.21)

Aromatic aldehydes and ketones can be reduced smoothly at room temperature by sodium dithionite in alkaline solutions using methyl viologen (MV^{2+}) as an electron transfer catalyst to obtain the corresponding alcohols in good yields [413]. Methyl viologen acted catalytically and assumed active species in the reduction were quinoid, MV^0 forms which were obtained by two-electron reduction of viologen. It should be noted, however, that reaction at room temperature proceeded very slowly (48 h). But it is well-known that in alkaline solutions aromatic aldehydes undergo the Cannizzaro reaction [414] to produce corresponding alcohol and acid, but authors did not consider this possibility. Besides, the ability of dithionite to reduce viologen to chinoid (neutral) form is also in question [415].

Louis-Andre and Gelbard have exclusively reduced α, β-unsaturated monoterpenic aldehyde (citral — mixture of geranial and neral) and ketones (isophorone, carvone, piperitone and pulegone) to the corresponding saturated aldehyde and ketones by the use of sodium dithionite under phase-transfer catalysis conditions [416]. Neither allylic nor saturated alcohols are formed. These results were confirmed later by Dhillon and coworkers [417]. In contrast to the results mentioned earlier, the same group has shown that sodium dithionite at higher temperature than in the paper [416] (85°C vs. 50°C, H_2O/dioxane in both cases) reduces exclusively the carbonyl group of functionalized aldehydes and ketones to the corresponding functionalized alcohols in good yield [418]. The other functional groups such as amides, esters, C=C double bond, halogens, alcohols, phenols, ethers, etc. when present along with carbonyl groups are not effected by this reagent (authors tried almost exclusively different derivatives of benzaldehyde).

Camps and coworkers used sodium dithionite for reduction of methylcyclohexanones [419], conjugated dienoic carboxylic acids and esters [420], unsaturated conjugated ketones [421], and alkyl 2,4-alkadienoates [422] under conventional [419,420] and phase transfer [419–422] conditions. Reduction of methylcyclohexanones in benzene–water using Adogen as the phase transfer agent affords good yields of isomeric mixtures of the corresponding methylcyclohexanols [419]. Selective double bond reduction in unsaturated compounds is achieved in most cases [420–422], except reduction of hydrophilic ketones which leads to predominant formation of water soluble sulfur derivatives [421].

Sodium dithionite has been used also for reduction of steroidal ketones and enones [423], as well as benzils [424,425].

Benzils can be reduced by sodium hydroxymethanesulfinate in aqueous DMF at 100°C [426]:

$$\text{ArCOCOAr} \longrightarrow \text{ArCOCHOHAr,} \tag{6.22}$$

where Ar stands for Ph, 2-ClPh, 2-BrPh, 4-MePh and 4-MeOPh.

But, in contrast to reduction with dithionite, the formation of impurity was observed (15–20% of crude reaction product). This was isolated by chromatography and was shown to be the hydroxymethylbenzoin. This compound is known to be formed by the aldol condensation of benzoin with formaldehyde [426].

The reaction of sodium hydroxymethanesulfinate with benzaldehyde or aromatic aldehydes containing electron donating substituents in the 2 or 4-positions (4-MeOPh, 4-MePh) was shown to be very slow [426]. In contrast, aromatic aldehydes with electron withdrawing groups in the 2- or 4-positions (2-ClPh, 2-BrPh, 4-NCPh) were reduced within few hours in aqueous DMF at 100°C.

Harris has shown [427] that sodium hydroxymethanesulfinate efficiently dehalogenates phenacyl halides and other α-haloketones with formation of the corresponding ketones in a variety of solvents (ethanol, tetrahydrofuran, dimethylformamide or aqueous mixtures of these). Ethanol was used in most of the cases. Phenacyl halides are reduced slowly (24 h) in ethanol at room temperature or rapidly (<1 h) at reflux. Aliphatic α-haloketones were reduced more slowly (6–48 h) at reflux. But, as in case of reduction of benzils, during the reduction of 4-chlorophenacyl bromide a by-product was observed (20% of the reaction mixture). This impurity was shown to be the dimer (1). A similar impurity (2) was isolated from the reduction of 4-nitrophenacyl bromide (see Fig. 6.40).

Jarvis and coworkers [428] have suggested the following scheme of reaction between phenacyl chloride and sodium hydroxymethanesulfinate in DMF or water/DMF mixtures (see Fig. 6.41).

Enolate anion formation would explain why the yield of dione is dependent upon the amount of water present in the reaction, more water diverting the anion to give an increased yield of acetophenone. The small amount of phenacyl sulfone produced indicates that the phenacylsulfinate ion,

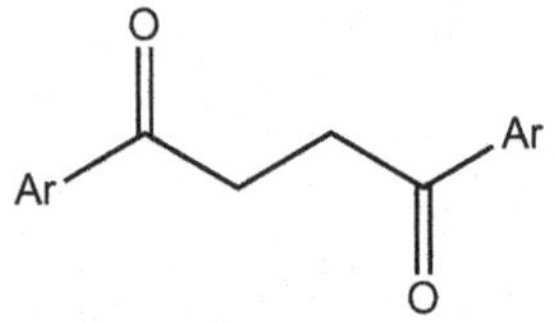

(1) Ar = 4-ClPh
(2) Ar = 4-NO$_2$Ph

Fig. 6.40   By-products of reduction of benzils and 4-chlorophenacyl bromide

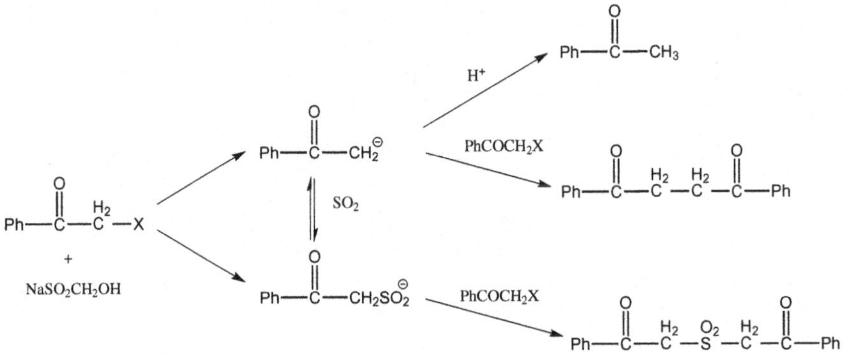

Fig. 6.41   Reaction of phenacyl chloride and sodium hydroxymethanesulfinate

PhCOCH$_2$SO$_2^-$, decomposes preferentially to the enolate anion rather than undergoing a bimolecular reaction with a molecule of phenacyl halide.

Nakagawa and Minami were the first who have studied in detail the action of thiourea dioxide on a variety of ketones and found that it is a convenient reducing agent for obtaining secondary alcohols from corresponding ketones [429]. Aliphatic, alicyclic, aromatic, and heteroaromatic ketones can be easily reduced with this reagent in good yield using an aqueous ethanolic solution in the presence of alkali. Authors assume that the reduction may be explainable in terms of a one electron transfer process.

Primary alcohols can be synthesized in the reaction between thiourea dioxide and such aldehydes as benzaldehyde, 2-pyridinecarbaldehyde, furfural, and 2-thiophenecarbaldehyde in ethanolic alkaline solution [430].

Dos Santos and coworkers described an efficient deoxygenation of α,β-epoxy ketones to enones, with thiourea dioxide in THF/alkaline medium

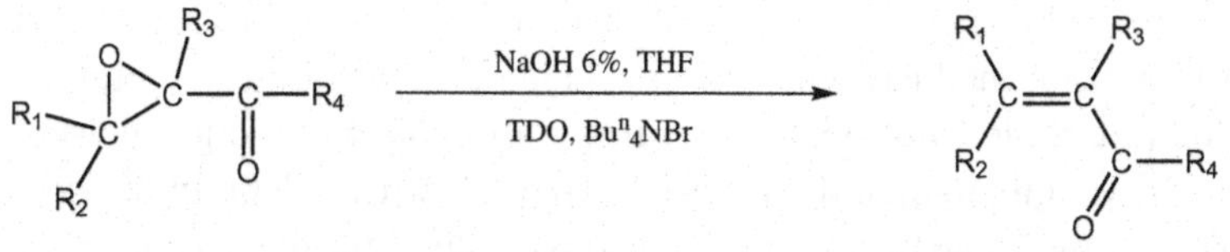

Fig. 6.42   Deoxygenation of α,β-epoxy ketones

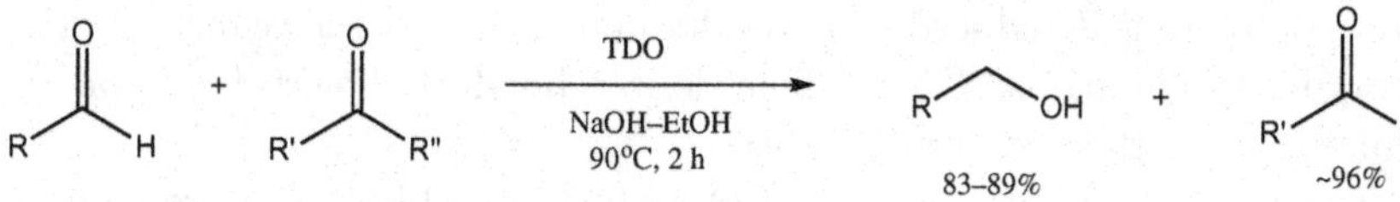

R = alkyl, aryl
R' = alkyl, aryl
R" = alkyl or aryl or Cl or OEt

Fig. 6.43   Chemoselective reduction of aldehydes using thiourea dioxide

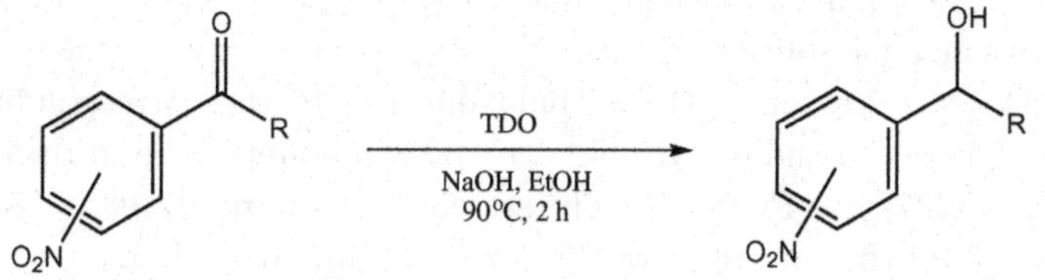

R = H, CH$_3$

Fig. 6.44   Chemoselective reduction of aromatic nitrocarbonyls to the corresponding nitroalcohols

and using tetra-n-butyl ammonium bromide as the phase transfer catalyst (see Fig. 6.42) [431].

Dhillon and coworkers have studied conventional [432,433] and microwave-assisted [434] chemoselective reduction of aldehydes [432–434] and ketones [433]. Reduction of a variety of aldehydes in presence of other carbonyl moieties to the corresponding alcohols occurs chemoselectively in high yields upon treatment with thiourea dioxide in aqueous alkali–ethanolic system [432] (see Fig. 6.43).

Aromatic nitroaldehydes and ketones were selectively reduced to the corresponding nitroalcohols by thiourea dioxide in aqueous alkali–ethanolic system in good to high yields [433] (see Fig. 6.44).

A series of papers were dedicated to reactions of TDO with ketones important from medical point of view [435–438]. Some representative oxosteroids (5α-cholestan-3-one, 3β-hydroxy-5α-cholestan-6-one, cholest-4-en-3-one, 5α-pregnane-3,20-dione) were treated with TDO in presence of strong alkaline reagent sodium n-propoxide in n-propyl alcohol (but not sodium hydroxide) [435]. The 3- and 6-ketones were reduced to alcohols; 20-ketones did not react. The fact that the reaction proceeds better in presence of a stronger base could indicate that the ketone may have to be enolized before it is reduced. The results indicated that the reducing agent favors back-side attack. The olefinic double bond of cholest-4-en-3-one is reduced before the oxo-group is affected.

Two years later, reactions of TDO with steroidal ketones were studied by Caputo and coworkers [436]. They have shown that thiourea dioxide does not play a major role in the reaction and, therefore, cannot be considered as an useful reducing agent for ketones since 89% of the overall yield of alcohols, originate from hydride attack by the alkoxide ion whereas only 11% of the total alcohols come from TDO reduction.

At the same time, Chatterjie and coworkers have reported that TDO in aqueous alkaline solution reduces N-substituted noroxymorphone derivatives such as naltrexone **(3)** and naloxone **(4)** to the corresponding 6β-hydroxy epimers (**5** and **6**), with no detectable amount of the corresponding 6α epimers [437] (see Fig. 6.45). The stereochemistry of these products was the opposite of that obtained in the corresponding hydride reductions.

Their next study revealed that TDO reduces the 6-keto group of morphine derivatives stereoselectively to the 6β-hydroxy epimers, and at the same time this reduction does not require the presence of 14-hydroxy group [438]. They have also found that, contrary to a report of Caputo *et al.* [436], thiourea dioxide reduces carbonyl groups (at least in the morphine series) in absence of alkoxide ions.

Note, in conclusion, that sodium dithionite and thiourea dioxide can be used for the reduction of the same aldehydes, mostly aromatic and heterocyclic. These aldehydes form unstable sulfinates which can be transformed to alcohols. There are no data on reduction of simplest aliphatic aldehydes, for example, formaldehyde, which forms stable sulfinate — hydroxymethanesulfinate. Rongalite reduces only aromatic aldehydes with electron withdrawing substituents. As for reduction of ketones, the best and most versatile reductant for them is sodium dithionite. The disadvantage of rongalite in some cases is the formation of by-products. Thiourea dioxide is

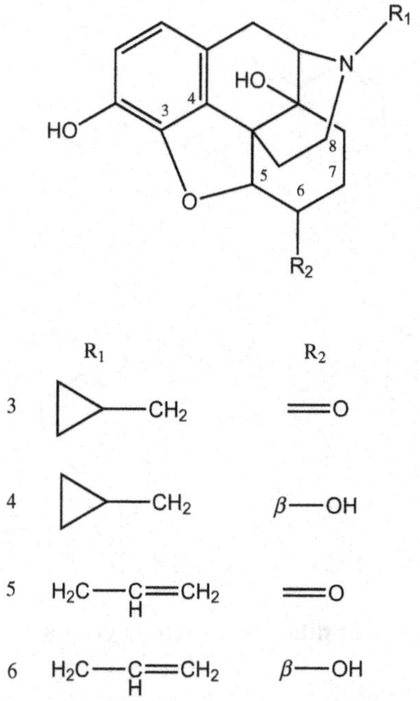

Fig. 6.45 Reduction of N-substituted noroxymorphone derivatives by TDO

capable of reducing variety of ketones but for some of them contradictory results have been reported.

6.5 Synthesis of Nitrogen-Containing Compounds

Thiourea dioxide and sodium dithionite are often used for reduction of nitrogen-containing compounds. In 1954, Gore briefly reported that aromatic nitro, azoxy, azo and hydrazo compounds can be successfully reduced by TDO to amines [439]. Reduction of nitrobenzene with TDO gave aniline and hydrazobenzene depending on the concentration of sodium hydroxide applied [440]. Azobenzene and azoxybenzene were found to offer hydrazobenzene as the final product by the same reagent [440]. Some other hydrazo derivatives were received from the corresponding azo compounds [441,442]. Chatterjie *et al.* have synthesized 2-aminomorphine and 2-aminocodeine by reduction of the corresponding 2-nitro derivatives with TDO [443].

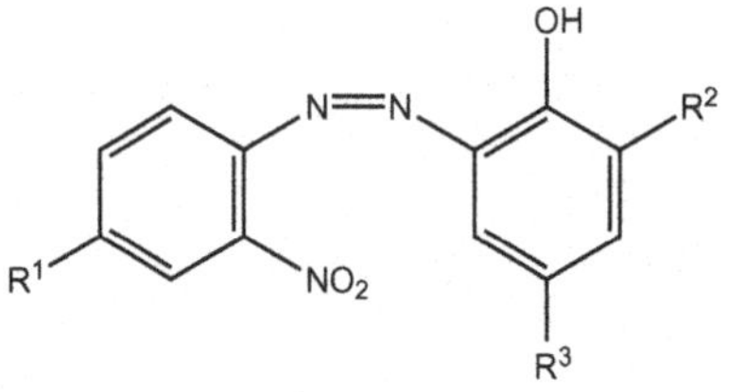

Fig. 6.46   Structure of different N-heterocycles synthesized by TDO

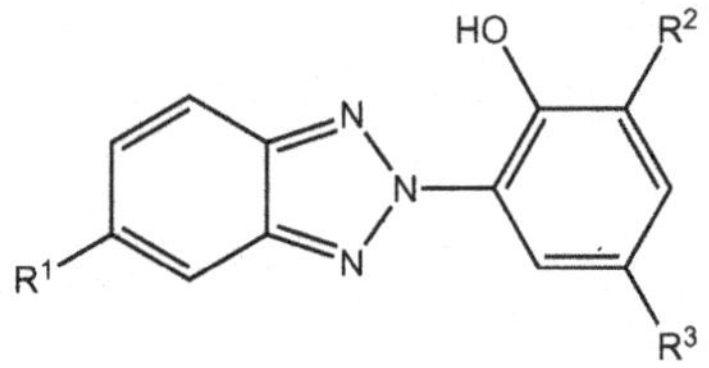

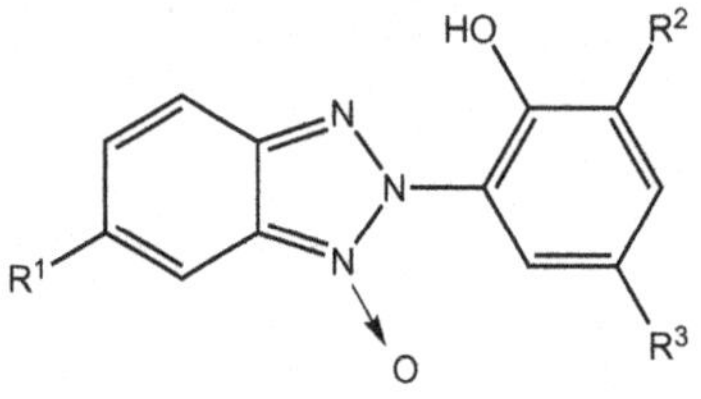

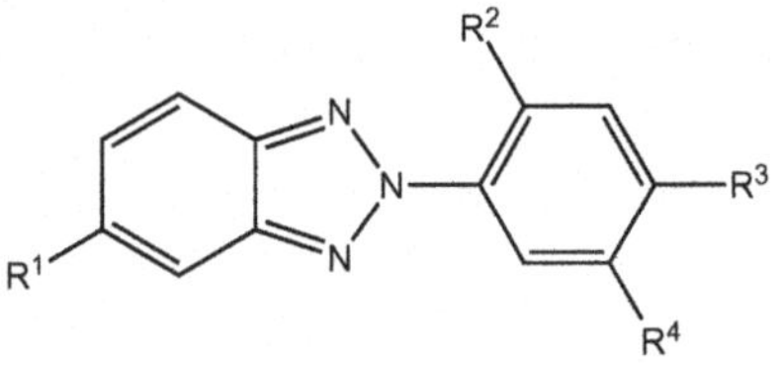

One of the main fields of applications of reactions between sulfur-containing reductants and nitro compounds is synthesis of N-heterocycles. Thus, thiourea dioxide reacts with o-nitrophenylazo dyes (1) in ethanolic alkali to give, depending on the reaction conditions, either the corresponding 2-aryl-2H-benzotriazoles (2) or their 1-oxides (3) [444] (see Fig. 6.46). The reagent is particularly effective for preparing high yield 2-(2′H-benzotriazol-2′-yl)phenols, which are important ultraviolet absorbers. Sodium dithionite was a less effective reducing reagent than thiourea dioxide. Nevertheless, conditions were established whereby sodium dithionite furnished the 2-(2′H-benzotriazol-2′-yl)phenols in high yield and purity.

Substituted o-nitroazobenzenes are readily reduced with thiourea dioxide to the corresponding 2-(aryl)-2 H-benzotriazoles (4), where R^1 = H, Me or Cl, R^4 = H and either R^2 or R^3 = OMe (and conversely R^3 or R^2 = H), or R^2 = R^3 = OMe [445]. Tanimoto and Kamano have published their independent work on the same reaction in a mixture of isopropyl alcohol, alkali and water at reflux [446]. Later the reaction of thiourea dioxide in ethanolic alkali with 2,2′-dinitrobiphenyl (5a) and several related dinitro compounds (5b) possessing an X-bridge (where X = NH, NMe, O and S) (see Fig. 6.47), located at the 1,1′-positions, at 85–90°C has been investigated by Wilshire [447]. With 2,2′-dinitrobiphenyl (which gives, depending on the

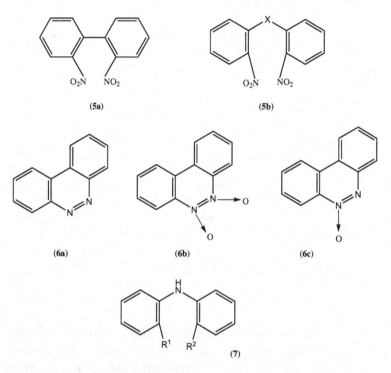

Fig. 6.47 Reaction of thiourea dioxide in ethanolic alkali with 2,2'-dinitrobiphenyl and several related dinitro compounds

amount of TDO used, good yields of benzo[c]cinnoline **6a** and its oxides, **6b, 6c**) and, to a minor extent, with 2,2'-dinitrodiphenylamine, an intramolecular reaction occurred to give heterocyclic products. With the most reducing agents, the reduction stops at the 5-oxide (**6c**) stage. Thiourea dioxide, therefore, has a distinct advantage over all the other reagents previously used for these transformations, except for lithium aluminium hydride, which is reported to reduce 2,2'-dinitrobiphenyl (**5a**) to benzo[c]cinnoline (**6a**) in 92% yield. With each of the other dinitro compounds, the only product obtained was formed as the result of a Smiles rearrangement (for a review of this rearrangement see Ref. [448]). Thus, the reaction of thiourea dioxide (4 equiv.) with N-methyl-2,2'-dinitrodiphenylamine gave 2-methylamino-2'-nitrodiphenylamine (**7**, R^1 = NHMe, R^2 = NO_2) in 72% yield.

Sodium hydroxymethanesulfinate can be also used for the synthesis of 2-aryl-2H-benzotriazoles (**2**) from nitrophenylazo dyes (**1**) in alkaline solutions at high temperatures (around 100°C) [449].

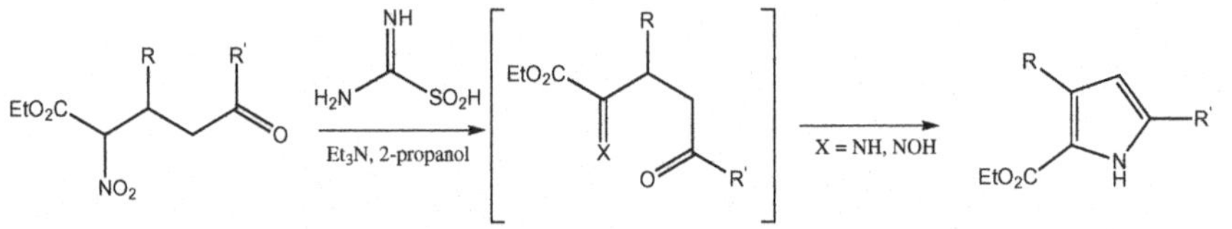

Fig. 6.48   Synthesis of pyrroles

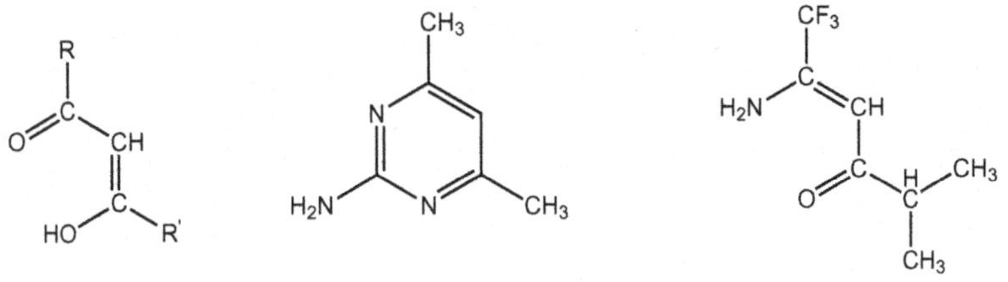

(8)   (9)   (10)

Fig. 6.49   Diones (8) and products of their reactions with TDO (9,10)

Quiclet-Sire and coworkers have found that γ-nitroketones with an electron-withdrawing group geminal to the nitro moiety can be reduced by thiourea dioxide to give ultimately the corresponding pyrroles as shown in Fig. 6.48 [450].

The desired reaction occurs by mere heating of the γ-nitroketone with the reducing agent in isopropanol containing some triethylamine. A number of pyrroles with various substituents were thus prepared in high yield.

Reaction of thiourea dioxide with 2,4-pentanedione (8, R = R′ = CH$_3$) gives 2-amino-4,6-dimethylpyrimidine (9) [451]. The reaction of TDO with 1,1,1-trifluoro-5-methyl-2,4-hexanedione (8, R–CF$_3$, R′=CH(CH$_3$)$_2$) leads to 2-amino-1,1,1-trifluoro-5-methyl-2-hexene-4-one (10) (see Fig. 6.49).

Treatment of thiourea trioxides 11a–11c, 12b and 13 (see Fig. 6.50) with sodium azide in acetic acid led to the corresponding aminotetrazoles, 14 (see Fig. 6.51). Compound 12a did not give any product to be characterized as an aminotetrazole [334].

Nitrogen-containing heterocycle, melamine (15) can be produced simply in the course of decomposition of thiourea trioxide in the weakly alkaline media (K$_2$CO$_3$) [39]. The other heterocycle, 2-ureido-4,6-dimethylpyrimidine (16) as well as melamine are formed in the reaction of thiourea trioxide with 2,4-pentanedione in presence of K$_2$CO$_3$ [39] (see Fig. 6.51).

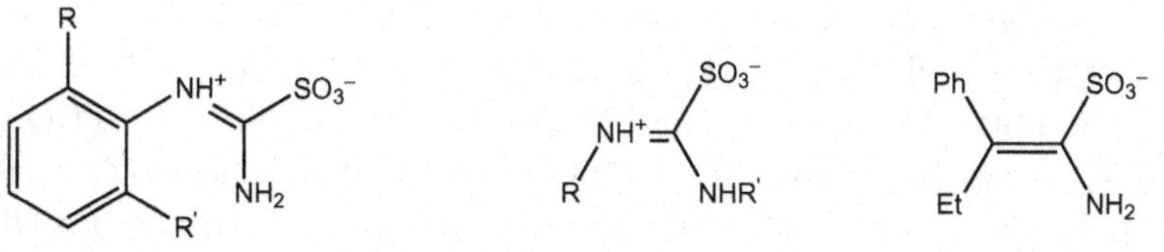

(11)   **(12)**   **(13)**

a R = R' = H; b R = H; R' = CH₃

c R = R' = CH₃

a R = R' = Et

b R = R' = Ph

Fig. 6.50   Thiourea trioxides used in reactions with azides

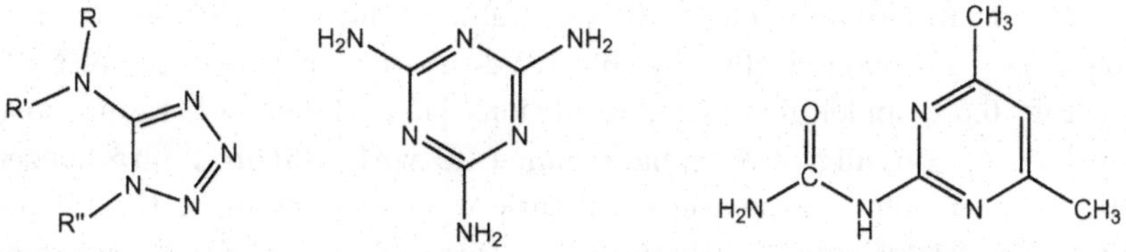

(14)   **(15)**   **(16)**

a R = H; R' = R" = Ph; b R = R' = H, R" = Ph
c R = R' = H, R" = 2-methylphenyl
d R= R' = H, R" = 2,6-dimethylphenyl
e R = Ph, R' = Et, R" = H

Fig. 6.51   Nitrogen-containing heterocycles produced from thiourea trioxides

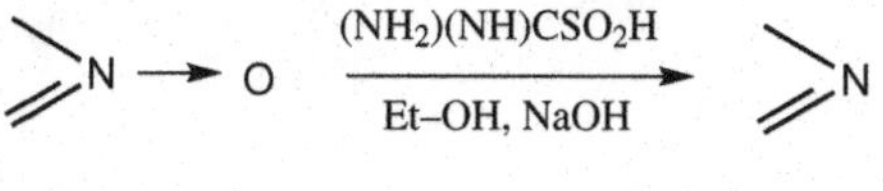

(17)   **(18)**

Fig. 6.52   Deoxygenation of N-oxides

Various heteroaromatic N-oxides **(17)** were efficiently deoxygenated to the corresponding bases **(18)** under mild conditions using thiourea dioxide [452] (see Fig. 6.52).

Reduction of nitro group by TDO is often used in analytical chemistry. Thus, reduction of aromatic nitro compounds by TDO results in formation of primary aromatic amines. Diazotization and coupling permit their spectrophotometric quantitation as the azo dyes. Conditions were established for nitrobenzene, nitrotoluenes, chloronitrobenzenes and toluenes, alkoxynitro derivatives, nitrobiphenyls, nitronaphthalene, dinitrobenzene, dinitrotoluene, dinitrochlorobenzene, and nitroacetanilides [453]. Later, TDO was applied for separate determination of p-, o-, m-nitrophenols [454], ethers of nitric acid, nitroso- and azocompounds [455], and nitrostilbenes [456]. Reaction of N,N'-dimethylthiourea with hydrogen peroxide, resulting in the dioxide, can be used to determine the content of H_2O_2 in biological systems [457].

Sodium dithionite often serves as the reagent for reduction of several nitrogen fuctions in organic compounds: nitro, nitroso, N-oxide, azo, azido, imine, oxime, pyridinium (numerous references can be found in the review [15]). Here, we consider just some of the most interesting results published after this review. Roy and Pramanik have synthesized pharmacologically and biologically valuable 1,2-disubstituted benzimidazoles (**24**, see Fig. 6.53) under metal-free conditions [458]. The synthetic approach involves (i) coupling of a primary amine **20** with 1-fluoro-2-nitrobenzene **19**, by nucleophilic aromatic substitution, (ii) reduction of the coupled nitroarene **21** by sodium dithionite, and (iii) cyclization of the corresponding diamine **22** using an aldehyde **23**.

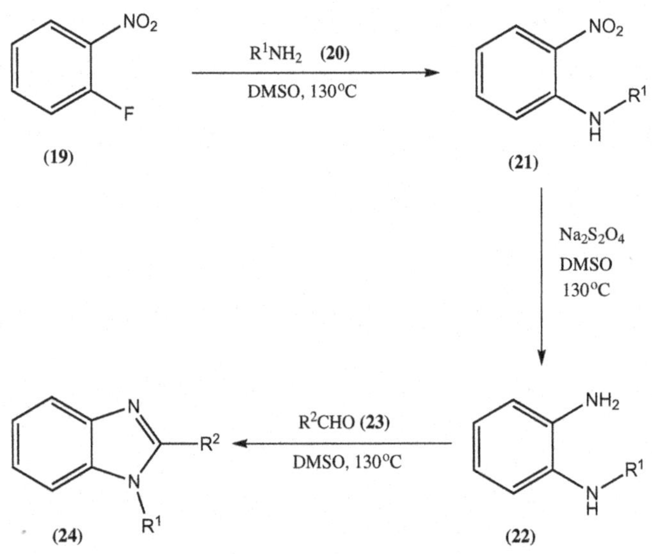

Fig. 6.53   Synthesis of 1,2-disubstituted benzimidazoles

A method for reducing aromatic nitro compounds on solid-phase supports using sodium dithionite has been presented by Scheuerman and Tumelty [459]. Conditions have been optimized to enable the use of this reagent for reductions on both polyethyleneglycol-polystyrene (PEG) resins and traditional polystyrene (PS) resins.

Reaction of nitroarenes with sodium dithionite in dichloromethane–water under phase-transfer catalysis (PTC) conditions has been studied by Kaplánek and Krchnak [460]. Tetrabutylammonium hydrogen sulfate was found to be an effective phase-transfer catalyst for this reaction. This method allows for the reduction of nitro groups to amino groups under mild conditions with complete conversion and is tolerant of other functional groups. This method is a superior alternative to tin(II) chloride-based reduction. Electron transfer catalysts viologens (1,1'-dialkyl-4,4'-bipyridiniums) are also utilized in the reductions of nitroarenes and nitroalkanes by sodium dithionite [461,462].

Reactions of viologens and sodium dithionite as such have been studied by many independent research groups [415,463–468]. In 1969, Carey *et al.* reported the consecutive formation of cation radical MV^+ (rapidly) and fully reduced methyl viologen MV^0 (slowly) by reduction of dication MV^{2+} with sodium dithionite in alkaline solution (Fig. 6.54) [463]. Later it was shown, however, that though during the reduction of an excess of dithionite the formation of an orange–brown species was observed, this product could not be oxidized quantitatively to the initial MV^{2+}. More reliably MV^0 has

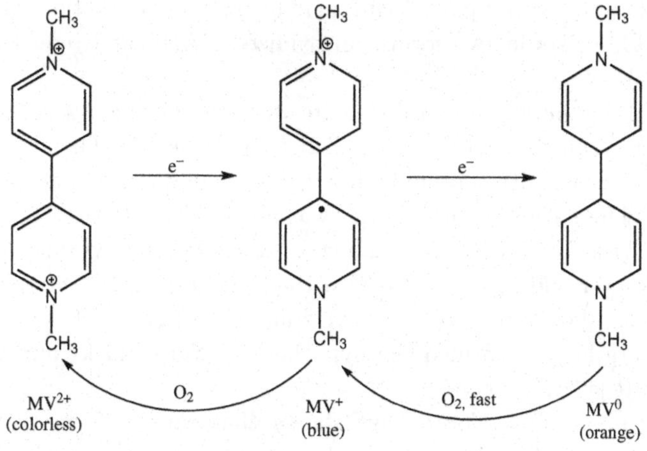

Fig. 6.54   Reduction of methyl viologen

been received with stronger reductant — sulfoxylate SO_2^{2-} formed from thiourea dioxide [415]. Introduction of oxygen to alkaline solutions of MV^0 in presence of an excess of both dithionite and TDO, leads to the formation of the cation radical MV^+, but further production of sulfoxylate during the decomposition of TDO converts MV^+ back to MV^0 at its initial concentration again.

Tsukahara and Wilkins have studied kinetics of reduction of eight viologens by dithionite [465]. The active reductant was the $\cdot SO_2^-$ anion radical, rate constants depend on the reduction potential of the viologen.

Reactions of dithionite with viologens were used as model processes in the investigations of microenvironmental effects on the kinetics of electron-transfer reactions [466,467]. One-electron reduction of dihexadecyl phosphate vesicle bound viologens by dithionite was studied by Thompson *et al.* [466]. Kim and coworkers have investigated the retarding effects of two polyelectrolytes on the rate of reduction of a series of dialkylviologens with alkyl chains varying from methyl to hexadecyl [467]. The origin of the retarding effect is shown to be dominated by electrostatic interactions in case of poly(styrenesulfonic acid) and by hydrophobic interactions in case of maleic acid-cetyl vinyl ether, which permits a true compartmentalization of viologen molecules. The latter polyelectrolyte was shown to be much more efficient than the former one.

Numerous papers are dedicated to the reductions of other derivatives of pyridine [469–476] including nicotinamide adenine dinucleotide NAD^+ analogs [469–472]. Nicotinamide adenine dinucleotide (25) is a ubiquitous biological molecule that participates in many metabolic reactions [477]. This dinucleotide exists in two forms, an oxidized NAD^+ and reduced NADH (Fig. 6.55) [478].

At high pH interaction of dithionite with NAD^+ analogs results in formation of a sulfinate adduct [469–472] (see Fig. 6.56). The same adduct has been received with sodium hydroxymethanesulfinate and thiourea dioxide [469]. The nature of this adduct was a subject of debate. Yarmolinsky and Colowick formulated the adduct as a 1,4-dihydropyridine derivative (26). Kosower and Bauer postulated that the adduct is a charge-transfer complex of sulfoxylate SO_2^{2-} and pyridinium ion (27) [470]. Cauchey and Schellenberg have confirmed the hypothesis of Yarmolinsky and Colowick using NMR [471].

Blankenhorn and Moore have shown that rate of formation of this adduct (reaction 1 of Fig. 6.57) is linearly dependent on dithionite concentration [472]. Hence, in their opinion, sulfur dioxide anion radicals do

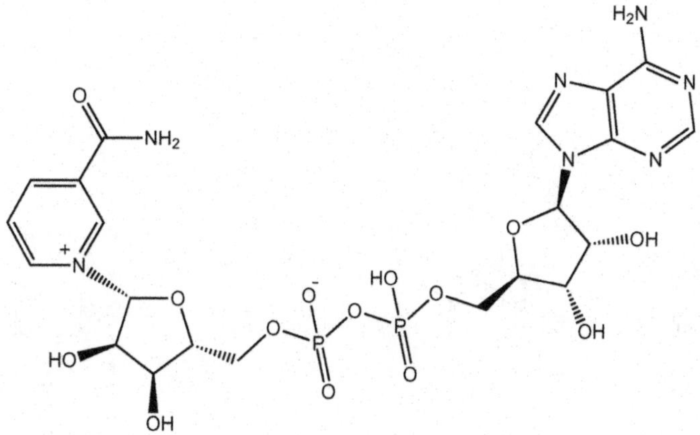

(25)

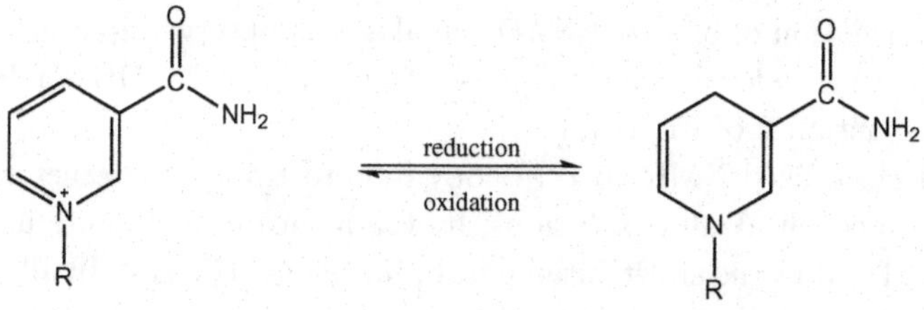

Fig. 6.55   Coupled redox reaction NAD$^+$/NADH

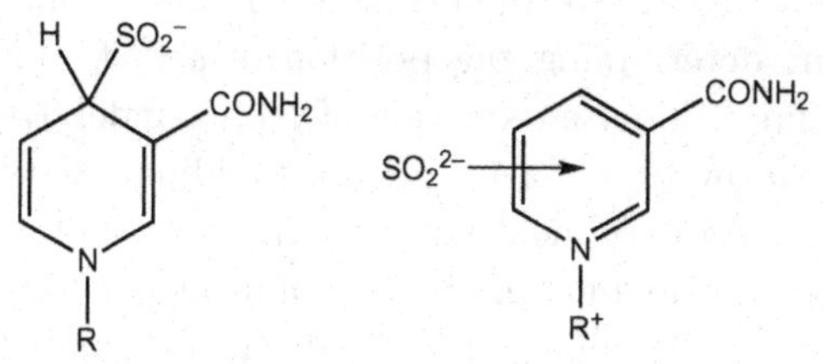

(26)   (27)

Fig. 6.56   Possible structures of sulfinate adduct

not appear to be involved in this process. The deprotonated adducts are found to be very stable both thermodynamically and kinetically. Formation of NADH analogs was therefore not observed at pH >11. Conversion of adducts, at high pH, to NADH analogs has been studied by pH jump studies, stopped-flow spectrophotometry: after protonation of the sulfinate

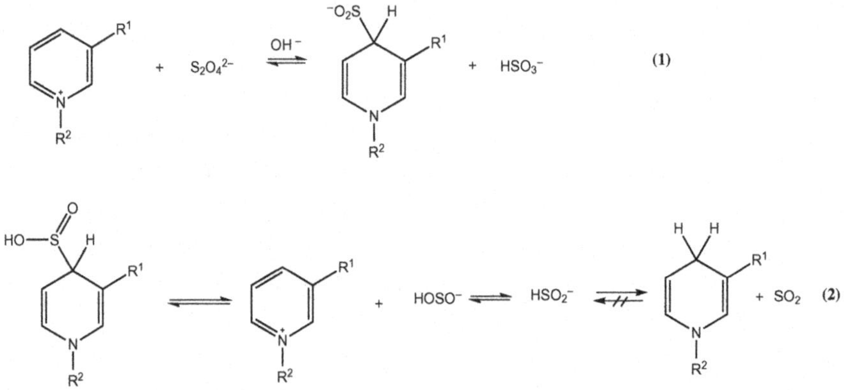

Fig. 6.57   Possible mechanism of the formation of NADH analogs

function, formation of oxidized NAD^+ analog was observed in a fast initial phase; in a much slower, second phase, formation of NADH analog was observed (reaction 2 of Fig 6.57).

Carelli *et al.* used NMR spectroscopy to detect and to characterize the adducts formed, in alkaline solutions, by the attack of dithionite anion on 3-carbamoyl or 3-cyano substituted pyridinium salts [473,474]. In all studied cases, only 1,4-dihydropyridine-4-sulfinates, formed by attack of dithionite oxyanion on the carbon 4 of pyridinium ring were found. This absolute regioselectivity seems to suggest a very specific interaction between the pyridinium cation and the dithionite through the formation of a rigidly oriented ion pair, determining the position of attack. Jokela *et al.* have shown that sodium dithionite reduction of 1-[2-(3-indolyl)ethyl]pyridinium salts leads to the formation of a 1,2-dihydropyridine derivative via the corresponding 1,4-dihydropyridine derivative [475]. A comprehensive review of the synthesis, mechanism and applications of nicotinamide cofactor analogs in redox chemistry, particularly 1,4-dihydronicotinamide derivatives and their oxidized counterpart, has been published recently by Paul *et al.* [476]. The focus of this review is placed mainly on the scope and limitations of these synthetic analogs in biocatalysis.

6.6   Organocatalytic Reactions

The term "organocatalysis" describes the acceleration of chemical reactions through the addition of a substoichiometric quantity of an organic compound which does not contain a metal atom [479,480]. The development of the field of organocatalysis in the last years has been very spectacular [481–485]. Many typical transition-metal-mediated reactions can now be performed under metal-free conditions [479]. Most organocatalysts used currently are bifunctional, commonly with a Brönsted acid and a Lewis base centers [485]. These compounds activate both the donor and the acceptor, thus resulting in a considerable acceleration of the reaction rate. Through explicit hydrogen bonding interactions organocatalysis combines supramolecular recognition with chemical transformations in an environmentally benign fashion [486]. Since urea and thiourea provide two hydrogen-bond donor groups that point in the same direction and which are spaced appropriately to interact with a range of anionic substrates, these functional groups are among the most popular binding motifs being used to prepare neutral anion-binding receptors [41,487]. The double hydrogen bonding interaction enables ureas and thioureas to interact with structurally diverse acceptors. The usage of urea or thiourea provides a method for altering hydrogen-bond-donating ability, while variation of the nitrogen substituents permits a high degree of fine-tuning of catalyst steric and electronic properties. The urea and thiourea structure is very convenient for the preparation of bifunctional catalysts, using amine coupling partners incorporating additional acidic or basic groups [488]. Owing to these properties, thiourea derivatives have become a subject of considerable interest for catalyst design in recent years [42,489–492]. It was shown that thiourea-based organocatalysts can successfully be used in different important C–C bond formation reactions.

Thiourea dioxide, owing to its easy accessibility, stability and strong hydrogen bonding ability, has a great potential as a promising organocatalyst [16–22,493–496]. Furthermore, TDO can provide higher levels of activation than thiourea because it has two oxygen atoms. Thiourea dioxide is almost insoluble in most common organic solvents and can therefore be recovered at the end of reaction and recycled. Since organocatalysis has emerged as a promising synthetic tool for constructing complex heterocyclic compounds via C–C, C–N, and C–O bond forming multicomponent coupling reactions, most of studies where TDO was used as an organocatalyst were dedicated to the synthesis of heterocycles. Interestingly, first TDO was used in organocatalytic reactions not only as such but in the host-guest complex with polyethylene glycol (PEG) [17]. This complex was found to be a very active catalyst for the direct synthesis of 3,4-dihydropyrimidinones with excellent yields (89–98%) via Biginelli condensation (Fig. 6.58). TDO

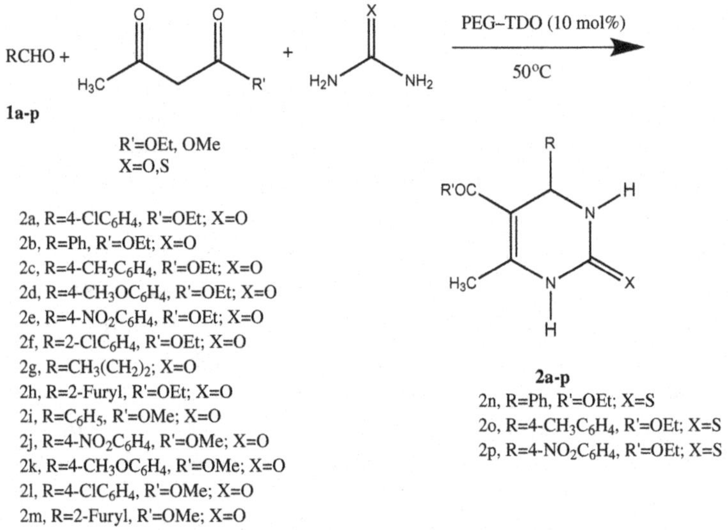
RCHO +
1a-p
H3C
R'
+
X
H2N
NH2
PEG–TDO (10 mol%)
50°C
R'=OEt, OMe
X=O,S
R
R'OC
H
N
H3C
N
X
H
2a-p
2a, R=4-ClC6H4, R'=OEt; X=O
2b, R=Ph, R'=OEt; X=O
2c, R=4-CH3C6H4, R'=OEt; X=O
2d, R=4-CH3OC6H4, R'=OEt; X=O
2e, R=4-NO2C6H4, R'=OEt; X=O
2f, R=2-ClC6H4, R'=OEt; X=O
2g, R=CH3(CH2)2; X=O
2h, R=2-Furyl, R'=OEt; X=O
2i, R=C6H5, R'=OMe; X=O
2j, R=4-NO2C6H4, R'=OMe; X=O
2k, R=4-CH3OC6H4, R'=OMe; X=O
2l, R=4-ClC6H4, R'=OMe; X=O
2m, R=2-Furyl, R'=OMe; X=O
2n, R=Ph, R'=OEt; X=S
2o, R=4-CH3C6H4, R'=OEt; X=S
2p, R=4-NO2C6H4, R'=OEt; X=S

Fig. 6.58 Biginelli condensation using PEG–TDO complex

itself catalyzes the reaction too, but the yield of product is slightly lower. After completion of the reaction, the catalyst could readily be recovered by precipitation with diethyl ether and reused for subsequent experiments (5 runs). Use of organic solvents, such as acetonitrile and ethanol did not enhance the reaction rates to any significant extent and, therefore, all the experiments were carried out under solvent-free conditions. Corresponding PEG–thiourea complex was found to be unreactive and no reaction occurred under similar reaction conditions.

The exact mechanism of the reaction is not clear; the probable mechanism of the reaction may involve the activation via the strong hydrogen bonding ability of the PEG–TDO complex with oxygen of the carbonyl group as shown in Fig. 6.59.

Two years later, the same group studied synthesis structurally diverse dihydropyrido[2,3-d]pyrimidine-2,4-diones (Fig. 6.60) in presence of thiourea dioxide [493]. Water was found to be an optimum reaction media for this transformation. The aqueous layer containing thiourea dioxide could be reused for several runs without significant loss in catalytic activity. No reaction occurred in the absence of catalyst under described reaction conditions.

Verma and coworkers have developed an efficient organocatalytic synthetic approach for the synthesis of a series of pharmacologically important

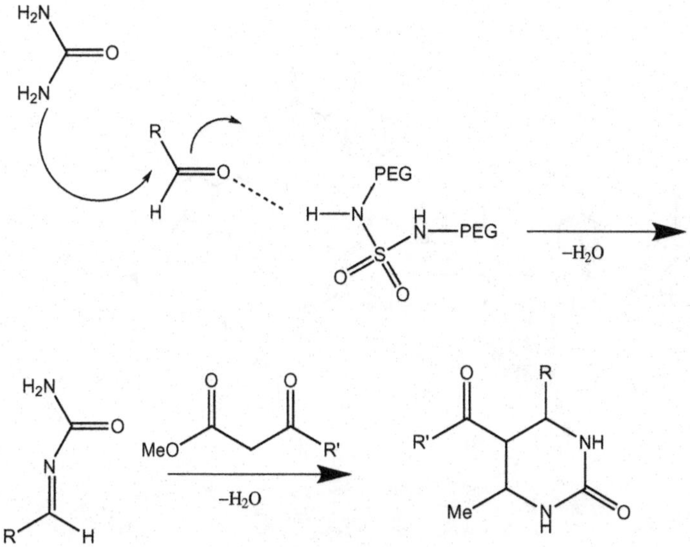

Fig. 6.59   Plausible mechanism for Biginelli condensation

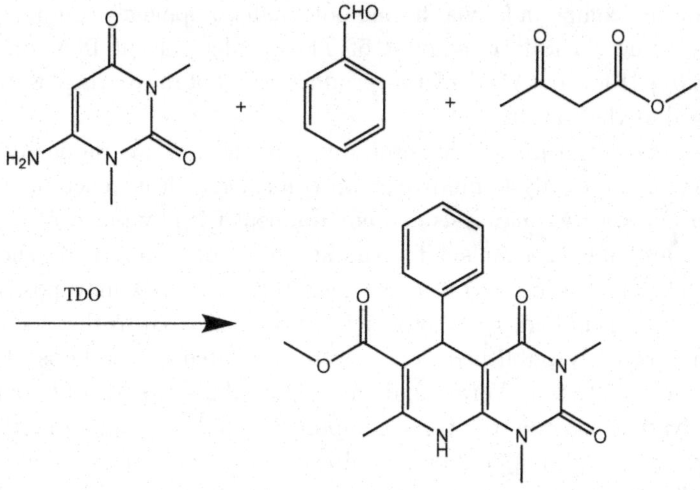

Fig. 6.60   Synthesis of dihydropyrido[2,3-d]pyrimidine-2,4-dione

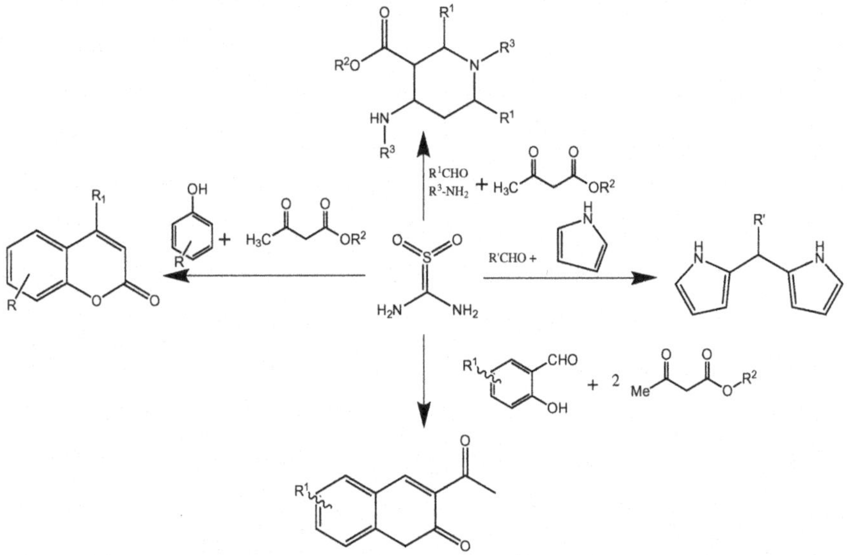

Fig. 6.61   TDO catalyzed one-pot synthesis of heterocycles

heterocyclic compounds via a one-pot multicomponent coupling reaction by using a catalytic amount of TDO under solvent free conditions (Fig. 6.61) [19]. Low yield of the product was obtained when water was employed as the solvent.

The salient features of this procedure are the easy synthesis and facile recovery of the catalyst, mild reaction conditions, clean reaction profiles, inexpensive starting materials and environment friendly protocol.

The probable mechanism for the synthesis of coumarin is shown in Fig. 6.62. The presence of two oxygen atoms in TDO may possibly be responsible for its higher catalytic activity as compared to the thiourea.

Three recently published papers were dedicated to synthesis of pyran derivatives [18,22,494]. Verma and Jain reported a synthesis of naphthopyran derivatives via one step three-component coupling of an aromatic aldehyde, α- or β-naphthol, and malononitrile catalyzed by thiourea dioxide in water (Fig. 6.63) [18]. The recovered aqueous solution of thiourea dioxide was recycled as such for next run.

A possible mechanism of the reaction is proposed in Fig. 6.64.

In contrast to other mechanisms of reactions catalyzed by TDO (see, for example, Figs. 6.59 and 6.62 for the syntheses under solvent-free conditions), authors assume that catalytically active form in the presence of water

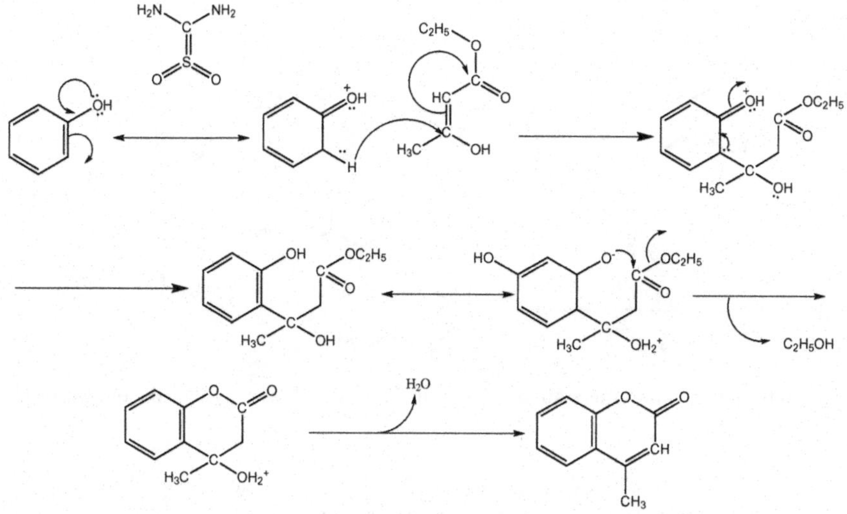

Fig. 6.62 Probable mechanism of synthesis of coumarin

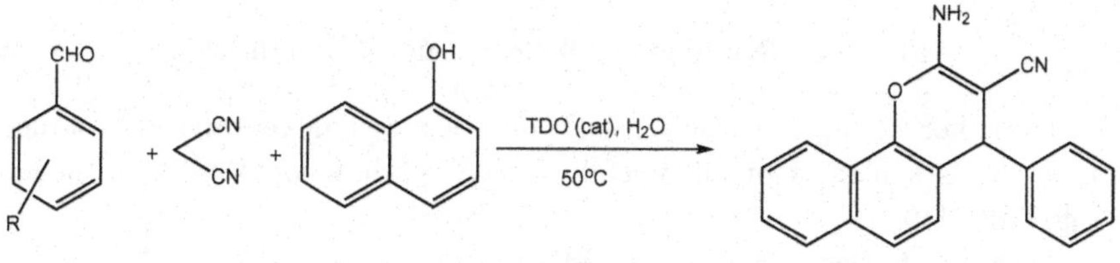

Fig. 6.63 Synthesis of naphthopyran

is aminoiminomethanesulfinic acid NH_2NHCSO_2H which can be formed after rearrangement of $(NH_2)_2CSO_2$. Thus, in formation of hydrogen bond with oxygen atom of substrate might participate not only NH-group, but OH-group as well, and solvent can strongly influence the mechanism of catalytic action of thiourea dioxide.

Aqueous thiourea dioxide was used as a catalyst in the synthesis of pyrano[4,3-b]pyran derivatives, [22] 3,4-dihydropyranol[c]chromenes and 6-amino-5-cyano-4-aryl-2-methyl-4H-pyrans [494]. As in the previously mentioned paper [493], a proposed mechanism assumes the activation of aromatic aldehyde by means of hydrogen bonding with aminoiminomethanesulfinic acid.

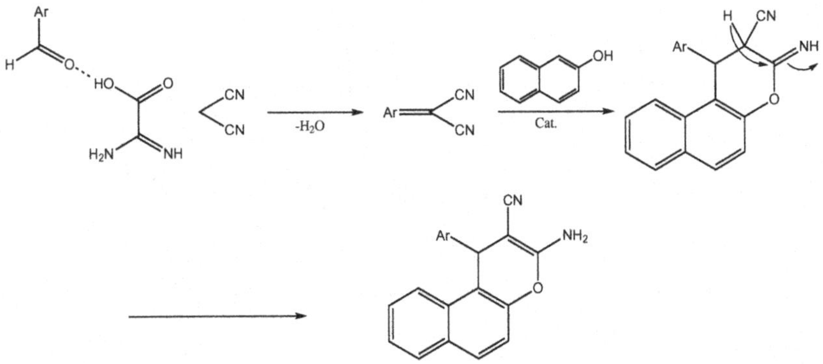

Fig. 6.64   Plausible mechanism for the TDO-catalyzed synthesis of naphtho-pyrans

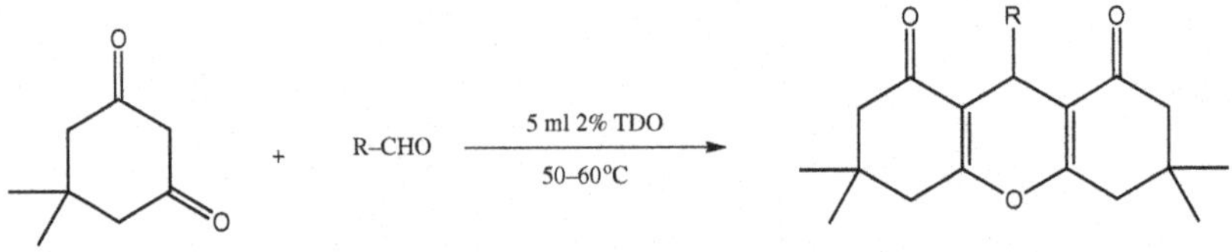

Fig. 6.65   Synthesis of 1,8-dioxo-octahydroxanthenes

The other example of reactions of aromatic aldehydes catalyzed by aqueous TDO is synthesis of biologically active 1,8-dioxo-octahydroxanthenes (Fig. 6.65) [495]:

Authors studied reactions of different types of aldehydes (including aromatic aldehydes possessing electron-releasing substituents, electron-withdrawing substituents and halogens on their aromatic rings) with dimedone in presence of thiourea dioxide as catalyst. In all these cases the desired products were obtained in good to excellent yields (82–97%). Though there are no mechanistic details in this chapter, one can assume that mechanism includes the activation of aromatic aldehyde by means of hydrogen bonding with aminoiminomethanesulfinic acid.

Aromatic aldehyde and dimedone as well as acetoacetate participate also in the one-pot synthesis of polyhydroquinoline derivatives catalyzed by aqueous thiourea dioxide (Fig. 6.66) [496].

Verma and coworkers studied also oxidation of sulfides [16] to sulfoxides with tert-butylhydroperoxide (TBHP) in presence of TDO (Fig. 6.67):

Usage of dichloromethane at room temperature was found to be optimum for this reaction in terms of reactivity and selectivity. Authors

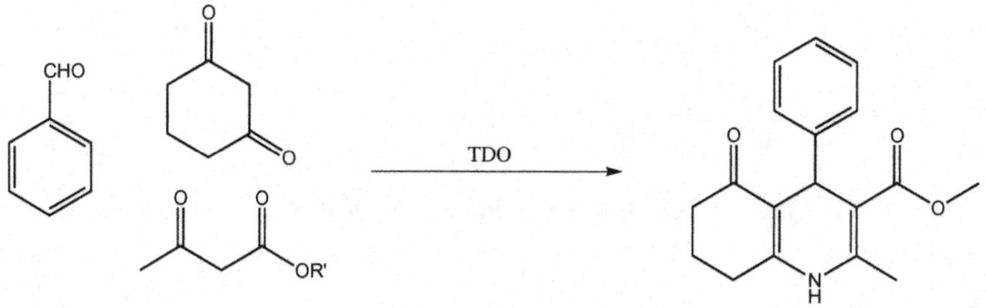

Fig. 6.66 Synthesis of polyhydroquinolines

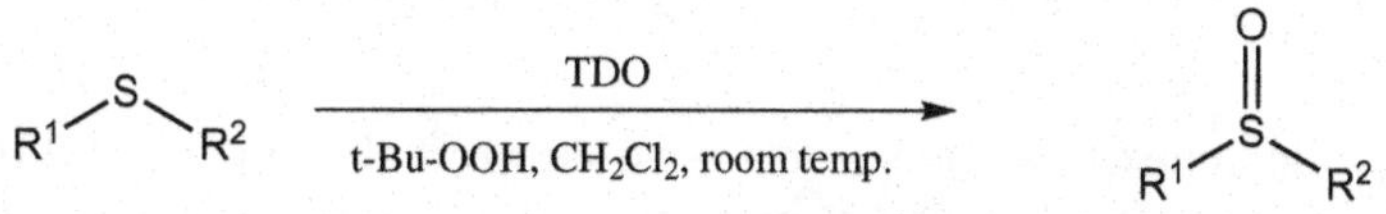

Fig. 6.67 TDO-catalyzed oxidation of sulfides to sulfoxides with TBHP

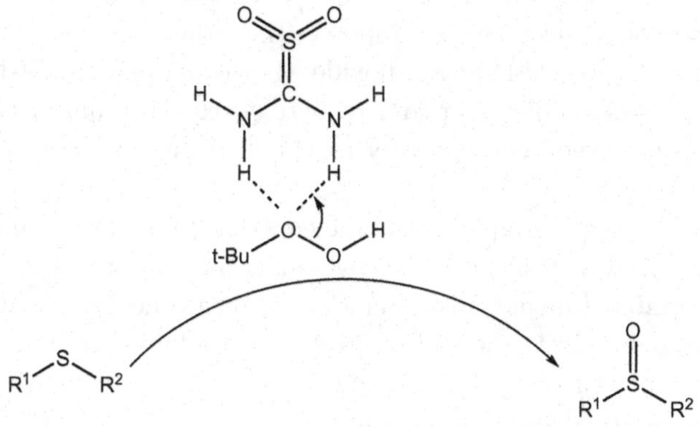

Fig. 6.68 Plausible oxidation pathway of oxidation of sulfides to sulfoxides in presence of TDO

established the higher catalytic efficiency of TDO as compared to thiourea, which is probably due to strong hydrogen bonding ability of TDO with TBHP.

The plausible mechanism of reaction is shown in Fig. 6.68. The higher efficiency of TDO than that of thiourea may be rationalized on the basis of strong hydrogen bonding interaction of thiourea dioxide with TBHP, which enhances the electrophilic nature of the oxygen of TBHP. Subsequently

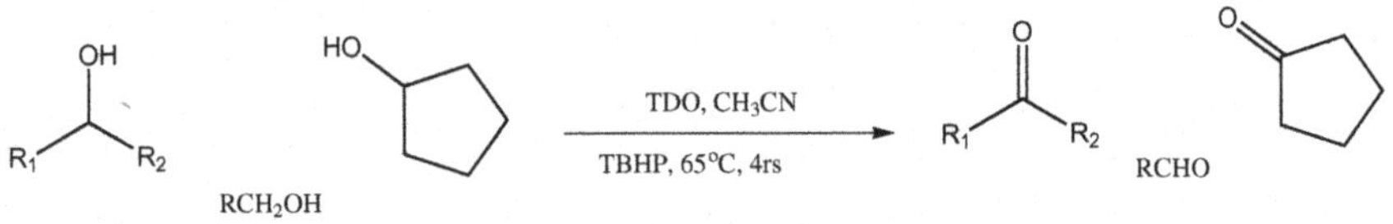

Fig. 6.69   TDO-catalyzed oxidation of alcohols with TBHP

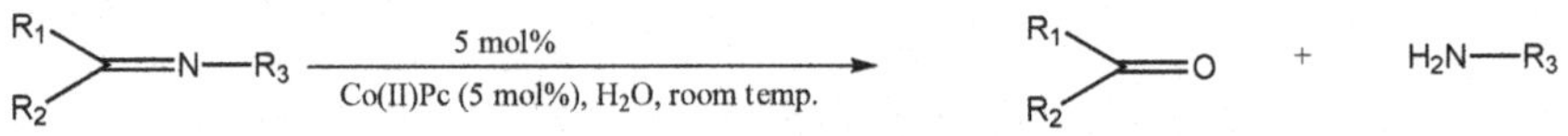

Fig. 6.70   Combined metal and organocatalyst promoted hydrolysis of imines

nucleophilic sulfur of sulfide reacts with electrophilic oxygen of TBHP to give corresponding sulfoxide. Note that in the Fig. 6.68 catalyst is shown as $(NH_2)_2CSO_2$ (non-polar medium).

TDO–TBHP system was used for the oxidation of alcohols (Fig. 6.69) [20]. Thiourea was found to be a less active catalyst. As in the case of oxidation of sulfides, it is proposed that the intermolecular hydrogen bonding between thiourea dioxide (Lewis base) and TBHP might increase the electrophilic character of peroxy oxygen atom of the TBHP. This activated oxygen subsequently reacts with the nucleophilic alcoholic group.

The interesting example of combined transition metal catalysis and organocatalysis has been considered by Kumar and coworkers [21]. In recent years, the combination of transition metal catalysts and organocatalysts has emerged as a new and powerful strategy for developing new and valuable reactions, and has attracted increasing attention as it can enable the development of unprecedented transformations that is not possible by use of either of the catalytic systems alone [497,498]. In particular, it was shown that combined catalysis by thiourea dioxide and Co(II) phthalocyanine (Co(II)Pc) leads to the fast hydrolysis of imines (Fig. 6.70) [21]: TDO or Co(II)Pc alone do not catalyze the hydrolysis of imines.

The probable mechanism involves dual activation of the C=N bond of imine and water molecules by the metal cation and thiourea dioxide through hydrogen bonding as shown in Fig. 6.71. This non-bonding association activates the water nucleophile toward attack on the imine bond, resulting in the cleavage of the imine to give the corresponding aldehyde and amine.

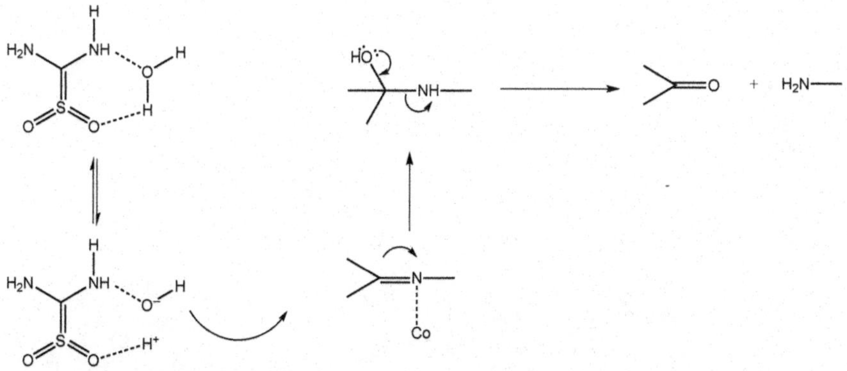

Fig. 6.71   Probable mechanism of catalytic hydrolysis of imines

In conclusion, one can say, that thiourea dioxide, owing to its strong hydrogen bonding ability, is an effective organocatalyst for different organic reactions. Di- and trioxides of other thioureas, not explored yet in organocatalysis, are promising organocatalysts as well.

Chapter 7

Inorganic Reactions
and Material Chemistry

7.1 Reduction of Graphene and Graphite Oxides

Graphene belongs to a new class of carbon nanomaterials — 2D materials made up entirely of conjugated sp^2 carbons arranged in a honeycomb structure [4,499]. These materials possess unique properties: enhanced electrical conductivity, high mechanical strength, high thermal conductivity, high impermeability to gases and optical transparency [4,499]. Graphene can be produced by micromechanical exfoliation of highly ordered pyrolytic graphite, epitaxial growth, chemical vapor deposition and the reduction of graphene oxide (GO) [500]. GO has two important characteristics: (1) it can be produced using inexpensive graphite as raw material by cost-effective chemical methods with a high yield, and (2) it is highly hydrophilic and can form stable aqueous colloids to facilitate the assembly of macroscopic structures by simple and cheap solution processes, both of which are important to the large-scale uses of graphene [500].

GO is generated from graphite oxide which was discovered more than 150 years ago. In 1859, British chemist Brodie added potassium chlorate to a slurry of graphite in the fuming nitric acid [501]. He determined that the resulting material was composed of carbon, hydrogen and oxygen. Later Staudenmaier added chlorate in multiple aliquots over the course of reaction as well as used additives of sulfuric acid to increase the acidity. Both methods allowed to achieve ratio C:O $\approx$ 2:1.4. Similar levels of oxidation have been achieved by Hummers and Offeman who used a mixture of potassium permanganate and concentrated sulfuric acid [501]. These three methods

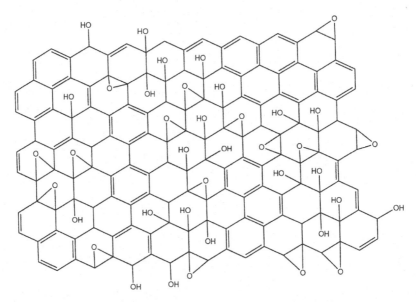

Fig. 7.1 Lerf–Klinowski model of graphite oxide with the omission of minor groups (carboxyl, carbonyl, ester, etc.) on the periphery of the carbon plane of the graphitic platelets of graphite oxide

comprise the primary routes for forming GO, and little about them have changed.

The most used material for oxidation is flake graphite, which is a naturally occurring mineral that is purified to remove hetero-atomic contamination [502]. The most widely cited in the contemporary literature model of graphite oxide is a non-stoichiometric, amorphous model of Lerf and Klinowski (Fig. 7.1) [501]. Based on this model, the oxygen functionalities encompass moieties such as hydroxyl, epoxide, carbonyl, ester and carboxyl groups. GO, being the exfoliated form of graphite oxide, is structurally different but chemically similar to graphite oxide. It retains the oxygen functionalities of its precursor, but largely exists as mono-, bi- or few-layer graphene sheets. GO is usually achieved via mechanical stirring or ultrasonication methods in a polar organic solvent or aqueous media [4]. To receive graphene from GO, different reduction methods are often applied. The pathway from graphite to graphene can be seen in Fig. 7.2.

Reduction of GO to graphene was first carried out using dimethylhydrazine at 80°C for 24 h [503]. Nowadays more than 50 types of reducing agents are used in the synthesis of graphene. Chua

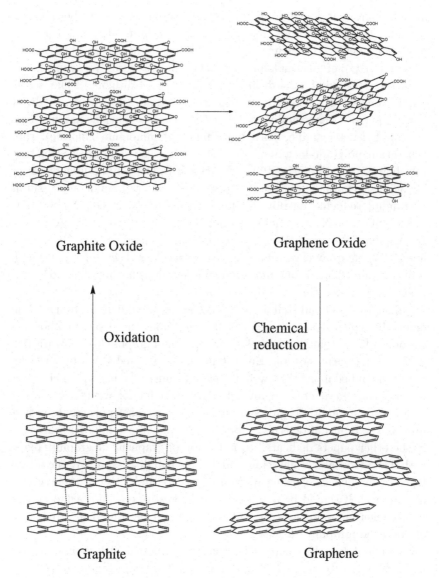

Fig. 7.2 Synthesis of graphene from graphite

and Pumera [4] suggested to distinguish two groups of reducing agents: (1) those which function according to well-supported mechanisms, and (2) those which function according to proposed mechanisms based on knowledge of organic chemistry. The first group includes borohydrides (sodium borohydride, sodium triacetoxyborohydride, sodium cyanoborohydride, ammonia borane), lithium aluminium hydride, hydrohalic acids (HI, HBr), sulfur-containing reducing agents (thiourea dioxide (TDO), ethanethiolaluminium chloride, Lawesson's reagent (2,4-bis (4-methoxyphenyl)-1,3,2,4-dithiadiphosphetane-2,4-dithione). The second group consists of nitrogen-containing reductants (hydrazine, hydroxylamine, pyrrole, benzylamine, p-phenylene diamine, ethylenediamine, urea, hexamethylenetetramine, poly(diallyldimethylammonium chloride), third generation poly(amido amine) dendrimer), oxygen-containing reductants (alcohols, namely MeOH, EtOH, iPrOH and BnOH), hydroquinone, ascorbic acid, saccharides (glucose, fructose, dextran, etc.), gallic acid), sulfur-containing reducing agents ($NaHSO_3$, Na_2SO_3, $Na_2S_2O_3$, $Na_2S \cdot 9H_2O$, $SOCl_2$, SO_2, $Na_2S_2O_4$, thiourea, thiophene), metal or metal salt (Al, Fe, Zn, Mg, $SnCl_2$) — acid (HCl, H_2SO_4) mixtures, metal (Al, Zn, Na) in alkaline (NH_3) environments, amino acids and peptides (L-cysteine, glycine, L-lysine, L-glutathione), plant extracts, microorganisms, proteins, hormones. It should be noted that this division into groups is rather arbitrary. Thus, surprisingly, TDO and sodium dithionite are in the different groups. Seemingly, this conclusion is based on the note from the paper of Zhou and coworkers [504], that the reduction mechanism of graphite oxide by dithionite is still an open question. But the mechanism of reduction by TDO has not been studied in detail as well. Chua and Pumera [4] just noted citing the paper of Nakagawa and Minami [429] that the reduction with thiourea dioxide in alkaline conditions proceeds via an electron transfer process to provide urea and sodium sulfite as end products. Mechanistic details are scarce also in the other papers dedicated to reduction of graphene or graphite oxides with TDO (see later). Therefore, further we will discuss mostly the advantages or drawbacks of using sodium dithionite and TDO in the synthesis of graphene. It should be noted that to the best of our knowledge, sodium hydroxymethanesulfinate (rongalite) has not yet been used as a reductant of graphene (graphite) oxides.

Zhou and coworkers noted [504] that the reduction with sodium dithionite is very fast (takes only few minutes at moderate temperatures), byproduct of reaction (Na_2SO_3) can be easily removed by water with no accompanying side-effect on the resultant graphene and its composites. It is also demonstrated that graphite oxide can be reduced with dithionite within the polymer matrix. After reduction, the conductivity of reduced graphite oxide is about seven orders of magnitude higher than that of graphite oxide and comparable to that of pristine graphite. The recovered electrical conductivity of reduced graphite oxide film indicates that the extensive conjugated sp^2-carbon network has been rebuilt in the reducing process.

Sodium dithionite was used for enhancing the absorbing capability of GO [505]. Acridine orange was the selected target to eliminate with GO as the adsorbent. Under identical conditions, GO without the *in situ* reduction showed a maximum adsorption capacity of 1.4 $g \cdot g^{-1}$, and GO with the *in situ* reduction provided a maximum adsorption capacity of 3.3 $g \cdot g^{-1}$. It was shown that sodium dithionite converts carbonyl groups on GO into hydroxyl groups, which function as the key sites for the adsorption enhancement.

Molina and his coworkers [506] have received polyester fabrics coated with reduced graphene oxide (RGO). In the reduction process they used sodium dithionite. The electrical conductivity was measured by electrochemical impedance spectroscopy (EIS) and showed a decrease of 5 orders of magnitude in the resistance when GO was reduced. Scanning electrochemical microscopy was employed to test the electroactivity of the different fabrics obtained. The sample coated with GO was not conductive since negative feedback was obtained. When GO was reduced to RGO the sample behaved like a conducting material since positive feedback was obtained.

In the last 4 years, TDO became one of the most important reductants of graphene (graphite) oxides [2–9,507]. The first study of reduction of GO with TDO in alkali has been performed by Chua, Ambrosi and Pumera [3]. In particular, they determined the C/O atomic ratio of GO before and after reaction since the increase of this parameter is an important evidence of GO reduction. Based on the X-ray photoelectron spectroscopy (XPS) analysis, which is a sensitive analytical technique in determining the surface elemental compositions and types of functional groups available on carbon materials (see Fig. 7.3), a C/O ratio as high as 16.0 was achieved between 2 h and 5 h of reaction. The C/O ratios of chemically reduced GOs (CRGOs) obtained in this study using thiourea dioxide were significantly higher than those obtained using other reducing agents such as sodium borohydride (C/O:5.8), hydrazine anhydrous (C/O:12.0), and hydrazine monohydrate (C/O:10.3). It was shown that reaction of GO with TDO leads to the reduction of C–O, C=O and O–C=O groups. The RGO product demonstrated a good electrochemical performance with a charge transfer resistance as low as 0.11 kΩ based on EIS measurements and a low overpotential for the oxidation of an important biomarker such as ascorbic acid. This indicates that reduced GO produced with TDO is suitable for electrochemical biosensing applications with improved performance.

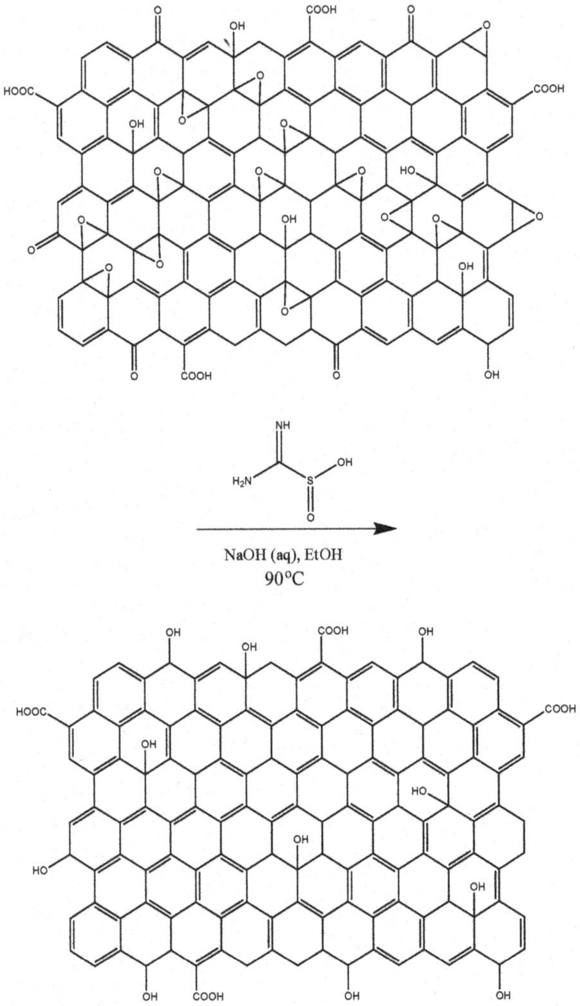

Fig. 7.3　Functional groups on GO reduced by TDO

Wang and coworkers have received high quality graphene from GO within 30 minutes with TDO as the reductant under moderate reaction conditions [2]. The C/O ratio of the TDO RGO was ca. 5.9 with the yield of graphene from GO > 99%. The relatively low value of C/O ratio in comparison with Chua's data [3] might be explained by much less reaction time.

Ma and coworkers [7] have received almost the same value of C/O ratio for TDO RGO (≈6) which is lower than for the material reduced by hydrazine hydrate (C/O:10.3) and hydroiodic acid reduction at 100°C (>12). However, graphene produced by TDO has a mean volume electrical conductivity, slightly higher than that of graphene obtained by hydrazine hydrate reduction, implying that π-conjugation of CRGO was further restored after TDO reduction. Moreover, compared with graphene produced by HI acid reduction at 100°C, this approach for the reduction of GO by TDO at room temperature will not produce toxic components, which is more safe and environment friendly. Surprisingly, though authors note that in alkaline solution TDO can produce sulfoxylate, they suppose to consider bisulfite (product of oxidation of sulfoxylate by dioxygen) as the active reducing species, but not much more active SO_2^{2-}. Unfortunately, they (as all other authors of the papers on reduction of GO with TDO) did not perform experiments under anaerobic conditions to exclude the influence of oxygen.

Wang and coworkers [9] have prepared reduced graphite oxide using TDO as reductant and polyvinylpyrrolidone (PVP) as stabilizer. It was shown that the reduction of graphite oxide could be readily achieved in 10 minutes, a reaction time which is much shorter than those required in common reduction reactions. The stabilizer, PVP, which could be easily absorbed onto the surface of reduced graphite oxide, provided material with good water and organic solvents solubility.

Carboxyl graphene was prepared on the scale of several grams through the reduction of GO by TDO at a relatively low temperature (40°C) in an alkaline medium (pH = 10) [6]. Characterizations of these materials demonstrated that selective reduction of the epoxy and carbonyl groups occurred on the surface of GO, while carboxyl groups were left behind. The presence of these remaining carboxyl groups may allow these materials to be modified further, thus extending the scope of applications for graphene. Carboxyl graphene samples had a higher C/O atomic ratio (4.4) than the graphite oxide samples (2.1), which were comparable with those observed among samples that were reduced by urea (4.5) [6]. A proposed mechanism for the selective reduction of GO by TDO at a low temperature under alkaline conditions is shown in Fig. 7.4. As can be seen from this figure, the active reducing species is sulfoxylate which reduces epoxy and carbonyl groups (it should be noted, however, that better to draw sulfoxylate as SO_2^{2-}, but not $^-S\text{-}(=O)O^-$).

TDO was used for preparation of three dimensional reduced graphene hydrogels (RGHs) with tunable pore sizes for electrode materials in supercapacitors [507]. By adding different amounts of TDO, the obtained

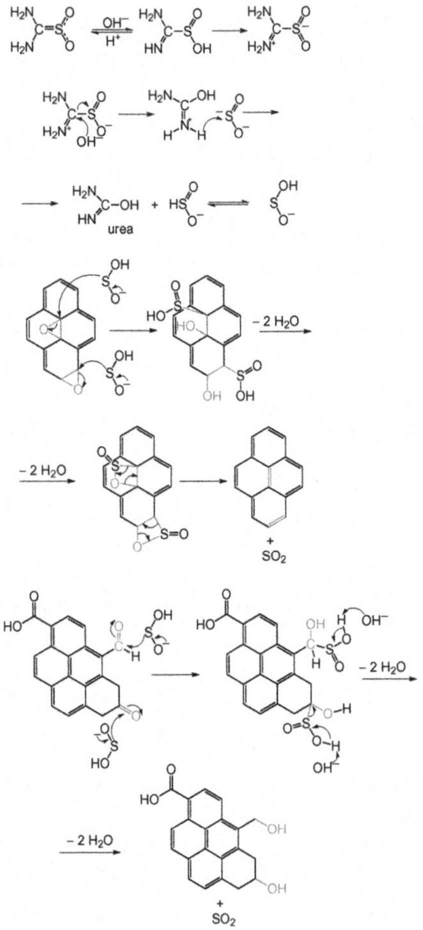

Fig. 7.4   Proposed mechanism for the reduction of GO by TDO in alkaline aqueous solution

RGHs behave in different degree of reduction, controlled specific surface area and pore size distribution, and unlike performances in supercapacitors. Interestingly, authors observed doping of nitrogen and to a lesser extent sulfur which positively influences the electrochemical properties of material.

RGO prepared by the reduction of GO using TDO was examined for the adsorption of Th(IV) from aqueous solutions [5]. The removal efficiency of Th(IV) increased sharply with an increase of pH from 1.0 to 5.0, and got a maximum value at pH = 6.0. The adsorption efficiency of Th(IV) was low at pH value below 3–4 due to the protonation of residual oxygen-containing groups on RGO (such as hydroxyl group) and the competition between H^+ and Th(IV) ions for the same adsorption site.

Park and coworkers [8] conducted reduction reactions of GO using three selected reductants (ethylene glycol, hydrazine and thiourea dioxide). With ethylene glycol reduction reaction did not sufficiently progress and as a result the conductivity of RGOs was observed to be relatively low. For RGOs made by hydrazine and TDO no significant differences in the degree of reduction, conductivity and dispersity in water was observed. However, RGO prepared by TDO showed an exceptionally good solubility in N-methylpyrrolidone, and the solution was stable for more than 4 months.

Thus, results of all recent studies show that TDO is an effective reductant of graphene (graphite) oxides and could be applied for industrial scale preparation of graphene.

7.2 Reduction of Metal Complexes.
Synthesis of Metals and Metal Sulfides

Reactions of sodium dithionite with metal complexes belong to the most well-studied systems from kinetic point of view. The results of the studies of these redox reactions are often used in textbooks on chemical kinetics. Especially important in this respect is the famous American textbook by Wilkins "Kinetics and Mechanism of Reactions of Transition Metal Complexes", in which these processes are discussed in detail [508]. The attention enjoyed by the kinetics of reductions by dithionite is due to the wealth of features that allow its reactions to illustrate the main themes of chemical kinetics.

Though dithionite is been used in the synthesis of reduced metal complexes for more than 50 years [509], the most important kinetic works which strongly influenced the chemistry of dithionite as a whole have been done approximately from 1973 to 1993, especially by the groups of Hambright and Wilkins [510–523]. In the first paper of Hambright's group [510], where authors studied reduction of manganese hematoporphyrin IX, it was shown that the rate constant linearly depends on the $[S_2O_4^{2-}]^{0.5}$ and $K_d^{0.5}$, where K_d is the equilibrium constant of reaction

$$S_2O_4^{2-} \rightleftharpoons 2 \cdot SO_2^-. \tag{7.1}$$

In their calculations they used a value for K_d equal to 6.225×10^{-10} M at 25°C [524]. In the same year Lambeth and Palmer [237] have published a slightly different value for this constant — 1.4×10^{-9} M — determined by EPR. This value was used in the most subsequent kinetic studies of reductions by dithionite.

Kinetics of reduction (to the divalent state) of a series of cobalt(III) and manganese(III) porphyrins (P) by dithionite in aqueous pyridine solutions [511]. The observed rate was independent of pH from 7.4 to 8.1 and half order in total dithionite. These results show that kinetically important reducing species is sulfur dioxide anion radical $\cdot SO_2^-$. The rates for Mn(III)P are ca. 100 times more than for those for corresponding Co(III) complexes. This can be explained in part by the greater ligand field stabilization energy of a formal d^6 Co^{3+} configuration vs. d^4 for Mn(III). In addition, bis(pyridine)manganese(III) hematoporphyrin is reduced at potentials 86 mV more positive than hematoporphyrin of (pyridine)$_2$ Co(III) in line with their relative reactivities toward dithionite.

Lambeth and Palmer [237] have investigated kinetics of the reduction by dithionite of several biochemically interesting substances (spinach ferredoxin, horse metmyoglobin, lumiflavin 3-acetate, horse heart ferricytochrome c, and spinach plastocyanin) under pseudo-first-order conditions (dithionite in excess). In case of ferredoxin, metmyoglobin and lumiflavin 3-acetate the observed pseudo-first-order rate constant (k_{obs}) linearly depended on the $[S_2O_4^{2-}]^{0.5}$, but reduction of cytochrome c and plastocyanin is described by the following equation:

$$k_{obs} = k_1[S_2O_4^{2-}]^{0.5} + k_2[S_2O_4^{2-}]. \qquad (7.2)$$

Thus, results with ferredoxin, metmyoglobin and lumiflavin 3-acetate support a mechanism involving $\cdot SO_2^-$, as the kinetically important reducing species, but cytochrome c and plastocyanin react with both $\cdot SO_2^-$ and $S_2O_4^{2-}$. When dithionite reacts with excess cytochrome c and lumiflavin 3-acetate, the reaction rate is virtually independent of the oxidant concentration and a limiting first-order rate constant of 1.7 s^{-1} is approached in each case. This rate constant has been assigned to monomerization rate of $S_2O_4^{2-}$, and its temperature dependence gives an activation energy of 24.1 kcal/mole for the dissociation of $S_2O_4^{2-}$.

Lambeth and Palmer have shown that kinetics of dithionite oxidation depends both on relative and absolute concentrations of reactants. As the dithionite concentration is increased to give pseudo-first-order conditions, the order of the reaction approaches half order in dithionite and first order in cytochrome c. Still higher concentrations do not affect the order with respect to cytochrome c but the order with respect to dithionite approaches

1. This is interpreted by the increased reactivity of $S_2O_4^{2-}$. Its role becomes increasingly important at higher concentrations (see Eq. (7.2) — the impact of the second term in the right side increases more effectively than the impact of the first one with an increase of dithionite concentration).

In many papers dedicated to reductions by dithionite is considered the interplay between outer (through a ligand)- and inner-sphere (which occurs subsequent to substitution for one ligand by the reducing agent) electron transfer as well as so-called bridging mechanism in which the reducing agent forms as an adduct with a coordinated ligand prior to electron transfer [525]. With the aim of clarifying some of the mechanistic details of dithionite reductions the study of the reductions of pentaamminecobalt (III) complexes $((H_3N)_5CoX)$ was undertaken [525]. These complexes are kinetically inert and have been studied with a wide variety of reducing agents that gave a sufficient basis for comparison. The observed pseudo-first-order rate constant has the general form (Eq. (7.2)) which is common for dithionite reductions. It was shown that the mechanism of reaction depends on the nature of X. The pyridine and ammonia complexes show only the k_1 term, and this is assigned to an outer-sphere mechanism for these as well as for the sulfato and chloro complexes. The second path (with $S_2O_4^{2-}$) occurs by a "bridging" mechanism for the azido, sulfato, trichloroacetato and benzoato complexes.

Marcus theory for outer-sphere reactions has been applied for the kinetics of reduction of several dicyanoporphyrinatoferrate(III) complexes to their Fe(II) forms by dithionite [512]. First of all, authors have confirmed in the previous results [510] that the rate constant linearly depends on the $[S_2O_4^{2-}]^{0.5}$ and $K_d^{0.5}$, i.e. sulfur dioxide anion radical is the reducing species and,

$$k_{\text{obs}} = k_1 K_d^{0.5} [S_2O_4^{2-}]^{0.5}. \tag{7.3}$$

The slope of the plot of k_{obs} vs. $[S_2O_4^{2-}]^{0.5}$ is thus considered to be equal to $k_1 K_d^{0.5}$, and with $K_d = 1.4 \times 10^{-9}$ M, the values of k_2 for each porphyrinato substrate can be obtained. The dicyanoporphyrinatoferrate(III) reduction rates are independent of cyanide concentration under conditions in which the diligated form predominates, indicating that the complexes have their full complement of ligands and are hexacoordinated in the activated complex. In the opinions of authors [512], this is consistent with an outer sphere assignment in their reactions with $\cdot SO_2^-$. As an extension of this work, Hambright and coworkers reported the kinetic behavior of the corresponding reductions of cyano cobalt(III) porphyrins [513]. The cyano cobalt(III) and iron(III) porphyrins behave in a remarkably different fashion. While the dicyano cobalt(III) porphyrins react with $\cdot SO_2^-$, the mono-cyano cobalt(III) form prefers $S_2O_4^{2-}$ as the reductant. Related work [526] has shown that hydroxy/aqua cobalt(III) porphyrins react with both $\cdot SO_2^-$ and $S_2O_4^{2-}$, whereas the bispyridine and di-aqua forms favor $\cdot SO_2^-$. Mehrotra and Wilkins reported on the reduction of 11 cobalt(III) complexes of a variety of types, as well as iron and manganese complexes [521]. About 8 of the 11 cobalt(III) complexes are reduced solely by $\cdot SO_2^-$ in the conditions used, and for the remaining three reduction by $\cdot SO_2^-$ may not be unduly sluggish, but this path is swamped by that involving direct $S_2O_4^{2-}$ reduction. On the basis of all these data (including results of Pinnell and Jordan mentioned above [525]) Hambright and coworkers have concluded that neither the structure nor the oxidation potential of the metal complex allows a prediction of its preference for $\cdot SO_2^-$, or $S_2O_4^{2-}$ as the favored reductant [513].

A series of papers has been dedicated to reactions of dithionite with proteins or peptides [237,515–520,527–531]. In most of the publications on the reductions of metalloproteins it was shown that the reaction involves the prior dissociation of $S_2O_4^{2-}$ to the radical $\cdot SO_2^-$, followed by one-electron reduction of the metal atom [515,516,518,520,527,528]. Thus, reduction of metmyoglobin (metMb) can be accommodated by the following scheme [520]:

$$metMb(H_2O) \; \underset{}{\overset{K_a}{\rightleftharpoons}} \; metMb(OH^-) \; + \; H^+, \tag{7.4}$$

$$metMb(H_2O) \; + \; \cdot SO_2^- \; \xrightarrow{k_{H_2O}} \; deoxyMb \; + \; SO_2, \tag{7.5}$$

$$metMb(OH^-) \; + \; \cdot SO_2^- \; \xrightarrow{k_{OH}} \; deoxyMb \; + \; SO_2. \tag{7.6}$$

The reduction of methemerythrin (Hr^+) by dithionite produces deoxy-hemerythrin (Hr^0) in multi, possibly three, stages [517]. The first stage is a fast reduction of methemerythrin to an intermediate A (combination of fully reduced to Fe(II) and non-reduced subunits) by $\cdot SO_2^-$ (it should be noted, however, that this stage of reduction of methemerythrin is slower than the reduction of myoglobin and hemin) [532]. The much slower second and third stages have rates independent of dithionite concentrations [517,519]. The nature of these stages are not clear. It can be the intermediate disproportionation within the octameric structure by utilizing the neighboring units, as seen in Fig. 7.5:

Another explanation for phase 2, unlikely it would appear, is that dithionite produces reduced centers in the protein during phase 1 and that

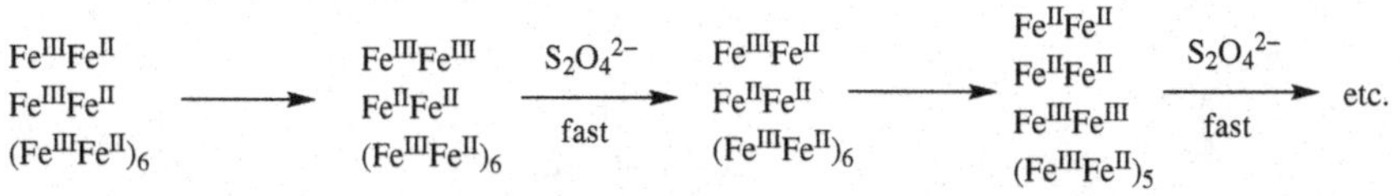

$$\begin{array}{c} Fe^{III}Fe^{II} \\ Fe^{III}Fe^{II} \\ (Fe^{III}Fe^{II})_6 \end{array} \longrightarrow \begin{array}{c} Fe^{III}Fe^{III} \\ Fe^{II}Fe^{II} \\ (Fe^{II}Fe^{II})_6 \end{array} \xrightarrow[\text{fast}]{S_2O_4{}^{2-}} \begin{array}{c} Fe^{III}Fe^{II} \\ Fe^{II}Fe^{II} \\ (Fe^{III}Fe^{II})_6 \end{array} \longrightarrow \begin{array}{c} Fe^{II}Fe^{II} \\ Fe^{II}Fe^{II} \\ Fe^{III}Fe^{III} \\ (Fe^{III}Fe^{II})_5 \end{array} \xrightarrow[\text{fast}]{S_2O_4{}^{2-}} \text{etc.}$$

Fig. 7.5   Possible consecutive pathway for reduction of methemerythrin

these act as intramolecular reducing agents towards the iron centers in the second phase. The third stage may be associated with the final reduction of remaining Fe(III) in the protein by some disproportionation process, a conformational change, or the breakdown of the oxy or dihydroxy bridge which is believed to exist in the met (Fe(III)), but not the deoxy, species. The slower phases after initial reduction of haem peptides by $\cdot SO_2^-$ has been observed also by Peterson and Wilson [529]. These interesting data show how complicated the reduction of some proteins can be.

In contrast to the most of metalloproteins, cobalticytochrome c is reduced by both forms of dithionite [530]. The rate of reduction with the dimer as reducing agent is much lower than with the monomeric radical (approximately 1,000 less, the same correlation has been received by Balahura and Johnson [533]). However, later these results have been questioned by Haim [531]. He noted that another explanation must be found for the observed biphasic behavior. Perhaps, there may be two protein species or cobalticytochrome c exists in two conformations.

Reduction of metal complexes by dithionite is often a part of catalytic processes [534–536]. Thus, overall mechanism of the reaction of sodium nitrite with sodium dithionite in the presence of cobalt(II) tetrasulfophthalocyanine, $Co(II)(TSPc)^{4-}$ in aqueous alkaline solution comprises the reduction of $Co(II)(TSPc)^{4-}$ by dithionite, followed by the formation of an intermediate complex between $Co(I)(TSPc)^{5-}$ and nitrite, which undergoes two parallel subsequent reactions with and without nitrite as a reagent [534]. Kinetic parameters for the different reaction steps of the catalytic process were determined. The rate of reduction of $Co(II)(TSPc)^{4-}$ was found to be independent of the $Co(II)(TSPc)^{4-}$ concentration. Rate constant (viz., $2.2\ s^{-1}$ at 25°C) are in close agreement with that reported for the dissociation of dithionite in Eq. (7.1), viz., $2.5\ s^{-1}$ (at pH = 6.5) [238], $1.8\ s^{-1}$ (0.1 M sodium hydroxide) [238], or $1.7\ s^{-1}$ (0.1 M sodium hydroxide) [237] at 25°C. This suggests that Eq. (7.1) is clearly the rate-determining step, which is followed by the rapid reduction of $Co(II)(TSPc)^{4-}$.

Cobalt tetrasulfophthalocyanine is also an effective catalyst for the reduction of nitrate by dithionite. The catalytic cycle includes reversible

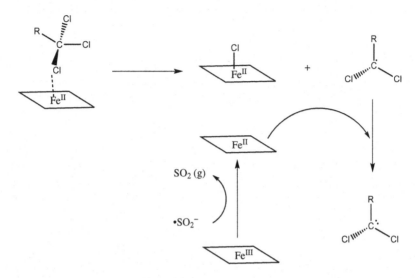

Fig. 7.6 Mechanism of reaction between heme and trichloromethanes in the presence of dithionite

reduction of Co(II) $\rightleftharpoons$ Co(I) and reduction of the coordinated substrate [534].

Larson and coworkers have shown that heme serves as an electron-transfer mediator in the dithionite-hematin-CCl_3R system ($R = NO_2$, $CHCl_2$, CN, Cl, COO^-, CH_3, H, $C(O)NH_2$ and CH_2OH [536]. Mechanism of this catalytic reaction is shown in Fig. 7.6.

Nzengung and coworkers have studied reductive dechlorination of per-chloroethylene (PCE) in homogeneous dithionite solution and at the surfaces of dithionite–citrate–bicarbonate treated ferruginous smectite and Na-montmorillonite [537]. It was shown that dithionite treatment of the Fe-poor Na-montmorillonite enhanced reductive dechlorination of PCE relative to dithionite-treated Fe-rich ferruginous smectite, i.e. the use of dithionite barriers for *in situ* treatment of chlorinated solvent plumes should not be limited to aquifers with Fe-rich sediments. Later Boparai and coworkers have shown that dithionite and dithionite-treated aquifer sediment and soil effectively dechlorinate chloroacetanilide herbicides [538].

The kinetics of reactions of metal complexes with TDO and sodium hydroxymethanesulfinate are studied lesser than with sodium dithionite. The reason is that in most cases the rate-determining step is the decomposition of these reductants but not reactions of reducing species with metal complexes. Just in a few papers reductions without any influence of slow decomposition of initial reductant is studied. The kinetics of reduction of iron tetrasulfophthalocyanine by sulfoxylate was studied in 0.1 M NaOH under pseudo-first-order conditions with an excess of sulfoxylate [535]. The reduction of the Fe(III) complex proceeds very quickly and could not be studied using stopped-flow techniques. The rates of both subsequent reactions (Fe(II)(TSPc)$^{4-}$ to Fe(I)(TSPc)$^{5-}$ and Fe(I)(TSPc)$^{5-}$ to Fe(I)(TSPc)$^{6-}$) depend linearly on the concentration of sulfoxylate (stable alkaline solution of sulfoxylate was received after complete decomposition of TDO under anaerobic conditions). The small negative value of $\Delta S^{\ddagger}$ and the large negative value of $\Delta V^{\ddagger}$ suggest an associative process (A or I_a mechanism) presumably due to inner sphere reduction of the Fe(II) complex, in which the associative attack of the strong SO_2^{2-} nucleophile on the electron-deficient Fe(II) center is the rate-determining step. The change in the iron redox state (from Fe(II) to formally Fe(I)) leads to drastic changes in the mechanism of reduction. On the basis of experimental data it was concluded that the most plausible mechanism for the reaction between the formally Fe(I) complex and sulfoxylate is an outer-sphere electron transfer from sulfoxylate to the tetrasulfophthalocyanine ligand. It was also shown that, contrary to dithionite, sulfoxylate is able to reduce formally Fe(I)(TSPc)$^{5-}$ to Fe(I)(TSPc•)$^{6-}$. This reaction is proposed to follow an outer-sphere electron-transfer process. Later EPR spectroscopy was employed to demonstrate chemical production of formally Fe(I) and Fe(0) states of phthalocyanines in the course of their reduction by sulfoxylate in water at room temperature, and physiologically-relevant pH [539]. However, reaction of sulfoxylate with several iron proteins (myoglobin, cytochrome c, rubredoxin) was found to reversibly yield Fe(II) with only little evidence for formally Fe(I) or Fe(0) states [540]. The reason why formally Fe(I) and Fe(0) states would be easily accessible in FeTSPc but less so in its biological cognates such as myoglobin may involve a carbon monoxide-generating side-reaction, and the fact, related to the size of the cavity within the macrocyclic ligand, that phthalocyanines have a lower affinity for carbon monoxide; arguments relating the size of the macrocycle to the energy of the e_g set of orbitals at the metal may also be proposed [539,540].

The kinetics of reduction of nitrite with TDO in presence of cobalt tetrasulfophthalocyanine was studied by Pogorelova and coworkers [541]. The process was shown to involve the Co(II) $\rightleftharpoons$ Co(I) catalytic cycle. The effective catalyst of nitrite reduction by TDO is also cobalt octasulfophenyl-tetrapyrazinoporphyrazine [542]. Cobalt and iron tetrasulfophthalocyanines catalyze also reaction of nitrite with sodium hydroxymethanesulfinate [543].

Dereven'kov and coworkers studied the reduction of μ-nitrido- and μ-oxo Fe-phthalocyanine dimer (μ-N(FeTSPc)$_2$ and μ-O(FeTSPc)$_2$, respectively) by sodium dithionite, TDO (sulfoxylate) [544]. The highest reduction rate was observed in case of sulfoxylate; a slower reaction proceeded when dithionite is used while the slowest reaction rates were observed using hydroxymethanesulfinate as the reductant. Production of formally Fe(0) state could be possible only with TDO.

A series of papers has been dedicated to reactions of sodium dithionite, TDO and sodium hydroxymethanesulfinate with cobalamins and their derivatives. Reaction of aquacobalamin (Co(III)) with sodium dithionite leads to the formation of unique, six-coordinate cobalamin(II) complex with sulfur dioxide anion radical $\cdot SO_2^-$ [545]. Later reduction of cyanocobalamin by dithionite and hydroxymethanesulfinate has been studied [546]. It was established that the character of the rate-determining step depends on the concentration of the reducing agents: when they are in excess, it is a step of elimination of dimethylbenzimidazole from cyanocobalamin; at lower concentrations of reductants a rate-determining is a step of their addition to cobalamin. Reaction of aqua/hydroxocobalamin with sulfoxylate in alkaline solutions leads to the formation of "super-reduced" cobalamin(I) [547]. This strong reductant is capable of reducing sulfur dioxide anion radical to sulfoxylate [548]. Data on kinetics and mechanisms of reactions between cobalamins and sulfur-containing compounds were recently reviewed by Dereven'kov and coworkers [549].

Sodium dithionite is used for the preparation [550,551] and extension reactive lifetime [552] of nano iron as well as synthesis of Fe/FeS nanoparticles [553]. Since the 1990s, the use of nano zero valent iron (nZVI) for groundwater remediation has been investigated for its potential to reduce subsurface contaminants such as polychlorinated biphenyls (PCBs), chlorinated solvents, and heavy metals [554–562]. Though the potential environmental risk of nZVI in *in situ* field scale applications are largely unknown [561], the contaminated water and soil treatment technology with nZVI has reached commercial status in many countries worldwide, however, is yet to gain universal acceptance [559]. nZVI can be synthesized by a number of methods, including the sonochemical method, the electrochemical method, the gas phase reduction method and the liquid phase reduction method. Among these, gas phase reduction and liquid phase reduction are the most common methods for synthesizing nZVI for remediation purposes [560].

The contradictory results of comparative studies of borohydride- and dithionite-synthesized iron-based nanoparticles have been published by two groups of researchers [550,551]. Sun and coworkers have shown that dithionite reduces Fe(II) and produces nZVI under conditions of high pH and in the absence of oxygen [550]. Though the nanoparticles are not pure iron, they effectively reduce trichloroethylene (TCE), not worse than nanoparticles synthesized with sodium borohydride. Later the similar study has been performed by Ma and coworkers [551]. The procedure for the synthesis with dithionite was similar to that described by Sun *et al.* [550] except that $FeCl_2$ solution was added dropwise into the sodium dithionite solution. The pH of the suspension was maintained at 9.5 or above. The particles formed using $NaBH_4$ (denoted $nFe(BH_4)$) principally contained Fe(0), while the particles synthesized using $Na_2S_2O_4$, (denoted $nFe(S_2O_4)$) were dominated by the mixed Fe(II)/Fe(III) mineral magnetite (Fe_3O_4) though with possible presence of Fe(0). The ability of both particles to reduce TCE under analogous conditions demonstrated remarkable differences with $nFe(BH_4)$ resulting in complete reduction of TCE while $nFe(S_2O_4)$ was unable to affect the complete reduction of TCE. However, considering the dramatic reduction in cost and the lack of hydrogen gas production (when use borohydride), a total cost-safety-benefit analysis would be required before the use of dithionite relative to borohydride is completely dismissed with fine-tuning of the synthetic process with dithionite [551].

Another reported use of dithionite involves the regeneration of Fe(0) from aged particles [552]. Sodium dithionite is able to effectively depassivate and restore the reducing capacity of oxidized nZVI particles to a similar or even more reduced state than fresh nZVI particles through a reaction with the surface passivation products [552].

Recently Kim and coworkers have developed a new one-pot method to prepare Fe/FeS nanoparticles using dithionite at room temperature [553]. Transition metal sulfides such as FeS are the effective reductants for chlorinated organic compounds in anaerobic environments, since several studies have found that the majority of the reduction capacity in natural waters is associated with sediment and aquifer solids, and this solid-phase reduction capacity has been attributed in part to Fe(II) and Mn(II) minerals and sulfide species [563,564]. However, multicomponent nanoparticles containing two or more different types of functionalities show unique physical and chemical properties, leading to significantly enhanced performance [553]. Thus, traditionally FeS is synthesized by the reaction between $FeCl_2$ and Na_2S [563]. In Kim's method, generation of Fe(0) and FeS (sulfide is formed in the course of decomposition of dithionite) via *in situ* precipitation occurs simultaneously in one-pot that offers advantages of simplicity, ease and rapidity. Resulting Fe/FeS nanoparticles have high surface area, good electrical conductivity and strong magnetic responsivity. In addition, the Fe/FeS shows a much higher reactivity toward contaminants, for example, TCE, than the pure Fe nanoparticles.

Jeong and Manthiram have described the synthesis of nickel sulfides by a reaction between nickel chloride and sodium dithionite in aqueous solutions at ambient temperature [11]. It was found that the compositions and structures of the products are controlled by the reaction pH and the amount of the reactants. While reactions under highly acidic (pH $\leq$ 2) and basic (pH $\geq$ 7) conditions yield crystalline sulfur and amorphous or poorly crystalline Ni_yS_x, respectively, those at intermediate $3 \leq pH \leq 6$ give crystalline Ni_yS_x. This solution-based synthesis procedure has accessed successfully the metastable spinel Ni_3S_4, which is otherwise difficult to obtain as a single-phase product by conventional high-temperature procedures. Note that according to Jeong and Mathiram, dithionite does not reduce Ni^{2+} at room temperature [11], while reduction to the zero-valent metal is possible at higher temperatures [565].

Rather unexpected results have been published by Khanna, who described one-step preparation method for the synthesis of copper and silver sulfides through the reduction of respective metal salts and sulfur by sodium hydroxymethanesulfinate [566]. Surprisingly, he has not received zero-valent metals, though copper and especially silver can be easily produced from respective salts by reduction of sodium hydroxymethanesulfinate [13,567–570]. The plausible reason is the different conditions used for the synthesis of sulfides and metals with hydroxymethanesulfinate. In the synthesis of silver or copper, Khanna *et al.* used water solutions and relatively low temperatures, for example, 60°C and room temperature in the synthesis of copper and silver nanoparticles, respectively [567,568]. For the synthesis of metal sulfides they used much higher temperature (130°C) and dimethylformamide as a solvent [566]. Obviously, in the last case decomposition of hydroxymethanesulfinate with formation of sulfide proceeds more rapidly, than reduction of metal ions. Unfortunately, mechanistic explanations of reductions of metal ions by hydroxymethanesulfinate are often unsatisfactory. Thus, Khanna *et al.* have performed reduction of Ag^+ by mixture formaldehyde-hydroxymethanesulfinate and assumed that Ag^+ is reduced by formate which is formed in the reaction of HCOH and HMS. However, direct reaction of Ag^+ (as well as Cu^{2+}) and HMS seems to be much more plausible. Indeed, Terskaya and coworkers have shown that Cu^{2+} can be reduced by HMS without any additional reductant [13]. In this study TDO was also used for the synthesis of copper nanoparticles.

Sodium hydroxymethanesulfinate can be used in the processes of chemical metallization of polyacrylonitrile (PAN) fibers [179]. The kinetics of reaction between Ni^{2+} and HMS has been studied by Egorova and coworkers [571] and Westbroek, Priniotakis and Kiekens [179]. In their book, authors have also described two-step process of metallization of PAN fibers (ab/adsorption and reduction of Ni^{2+} in one bath solution followed by galvanization) [179].

TDO is capable of reducing Cd^{2+} as well as Ni^{2+} [572,573]. Sotskaya and coworkers have examined the effect of Ni, Co and Cu nanoparticles synthesized with TDO on the characteristics of electrodeposited nickel coatings, silver-containing polymer matrix composites and porous silicon [574]. Sorption treatment of metal–polymer composites and porous silicon leads to considerable changes in surface morphology, accompanied by changes in the catalytic, electrical and optical properties of the material.

TDO can also be used in the synthesis of metal sulfides. Davies and coworkers have elaborated a method for the synthesis of ZnS for use in the preparation of phosphors for cathode ray tube (CRT) devices [10]. The preparation of ZnS by this method is extremely simple and does not yield large amounts of liquid or gas containing volatile sulfur species. Thus, this method has been shown to be an excellent method for the preparation of ZnS phosphors, particularly copper-activated materials, requiring no purification of the reagents, with little production of sulfur-containing waste species, and resulting in small particle size powders without postproduction milling or separation of the powders. These phosphors have been shown to have exceptional luminescent properties compared to standard commercial materials. Davies and his coworkers assumed that the reactive species from TDO is sulfur dioxide anion radical which is formed directly from TDO as a result of homolytic C–S bond cleavage (the same conclusion has been earlier done by McGill and Lindstrom [572]). But, as mentioned above, the primary product of TDO decomposition in water solutions is sulfoxylate, but not sulfur dioxide anion radical.

In the conclusion, note the application of TDO in the preparation of technetium labeled radiopharmaceuticals [575–578]. Technetium-99m as pertechnetate requires reduction as part of the process of binding the Tc-99m to various carriers in radiopharmaceuticals. Stannous chloride, the most commonly used reducing agent, is rapid and effective at room temperature. However, it is easily oxidized to stannic ion by oxygen and is rapidly hydrolyzed to a colloidal stannous hydroxide which effectively binds Tc-99m. In addition, residual stannous salts in the body may cause localization of pertechnetate in subsequent brain scans [575]. The results of study of Fritzberg and coworkers [575] indicate the potential utility of TDO as an alternative reducing agent to $SnCl_2$. In addition to producing clean binding of technetium to chelating agents, its greater stability makes it more convenient and the lack of colloid binding problems reduces the variability of radiopharmaceutical preparations. TDO was also effectively used in the preparation of Tc-99m glucoheptonate [576]. This product has high radiochemical purity and stability even at high levels of technetium-99m activity. Baldas and Pojer, however, found that preparations of ^{99m}Tc-HIDA (N,[α-2,6-dimethylcarbamoylmethyl]iminodiacetic acid) using TDO as reductant were significantly different from those using stannous chloride both in chromatographic behavior and *in vivo* distribution in mice [577]. These differences have been shown as a result of complex formation of TDO (or its decomposition products on heating) with reduced pertechnetate. Later Neves and coworkers described the results of labeling kinetic studies as well as radiochemical characterization and particle size evaluation of technetium-99m rhenium sulfide colloid using TDO as reducing agent [578]. Comparison with the same colloid which makes use of Sn-sodium pyrophosphate complex as reducing agent has shown higher labeling yields, simplification of labeling procedure and a longer shelf life when TDO is used.

It should be noted as well that very interesting and rather unexpected application of sodium dithionite has been reported recently. Its reducing properties are used in conservation, mainly for iron stain removal from both organic and inorganic substrates, for example, archeological wood, and occasionally to treat corroded copper and silver artifacts [579,580]. This application is based on the ability of dithionite to dissolve iron oxides (lepidocrocite, goethite) [579–583]. Natural goethite reduced with dithionite also has enhancing catalytic properties in the oxidation of some organic compounds [584].

The other examples of the use of compounds under consideration for reduction of metal-containing compounds can be found in the book authored by Davies *et al.* [584].

Reactions of sulfur-containing compounds with metallocomplexes are widely used in biochemistry and biochemical tests. Nowadays, $Na_2S_2O_4$ is one of the most widely used reductants in biochemistry, including reduction of proteins [585]. It is generally accepted that neither dithionite nor its monomeric form $\cdot SO_2^-$ affect protein structure, or the structure of active metal centers in metalloproteins in particular. Illustrating this issue, protocols for identifying types of hemoproteins in cell extracts rely precisely on reaction with excess dithionite [586]. Dithionite is commonly employed to generate the deoxy form in hemoglobin and other proteins [587,588]. Kinetics of reductions of hemoproteins (Fe(III) complexes) are very similar to other reductions with dithionite discussed in detail above; sulfur dioxide anion radical is the active reducing species (see, for example, Refs. [589] and [590]). Dithionite is often used in the different tests detecting sickle hemoglobin (HbS) [591,592], carboxyhemoglobin [593–595], microsomal cytochrome P-450 [596], hemoglobin [597,598] by visible spectrophotometry, gas chromatography and infrared microspectroscopy. Importantly, that dithionite tests for HbS are suitable for field screening. Principle of this test is, if the test sample contains HbS, it gets reduced by sodium dithionite. Reduced HbS is insoluble which makes the dithionite tube test buffer solution turbid. On the other hand normal hemoglobin gives clear solution, as it is not able to form turbidity [592].

Sodium dithionite is also employed in the studies of biologically and pharmacologically valuable dinitrosyl iron complexes (DNICs) with natural thiols — glutathione, cysteine, (RS) [599–601]. DNICs with thiolate ligands produce miscellaneous physiological and biochemical effects on animal and human organisms [599–602]. Their biological action mimics those of nitrogen monoxide (NO), an universal endogenous regulator of metabolic processes, and its oxidized form, viz., nitrosonium ion (NO^+), and is based on the ability of DNICs to act as NO and NO^+ donors in living systems [599–601]. At neutral, the diamagnetic binuclear form ($[(RS)_2Fe_2(NO)_4]$, B-DNIC) of DNIC was dominant (90%), while at alkaline pH, it was the paramagnetic mononuclear form ($[(RS)_2Fe(NO)_2]$, M-DNIC) that was dominant (90%). These findings can be due to the chemical equilibrium between the two forms, which, in its turn, depends on the concentration of glutathione ionized at thiol groups. After treatment with sodium dithionite, both M- and B-DNIC are converted into the same paramagnetic form with the same optical characteristics. Based on these data, it was suggested that sodium dithionite-reduced B-DNIC is subject to decompose reversibly into M-DNIC. Moreover, it was shown that protein bound B- and -DNIC with thiol-containing ligands possess redox activity commensurate to that of their low-molecular derivatives. Thus, experiments with sodium dithionite shed light on the mechanisms of the redox transformations of dominant at physiological media binuclear form of DNIC.

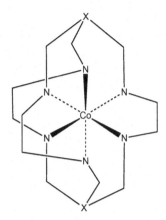

Fig. 7.7   Structure of bis(hydroxylamino)hexaaza cryptand

The interesting reactions of reduction of nitro groups in the cage complexes by sodium dithionite have been studied by Balahura and coworkers [603] as well as Lay and Sargeson [604]. The cobalt(III) complex of a dinitrohexaazacryptand, (1,8-dinitro-3,6,10,13,16,19-hexaazabicyclo[6.6.6] eicosane)cobalt(III), Co(dinosar)$^{3+}$, is reduced quantitatively by dithionite to the cobalt (III) complex of the corresponding bis(hydroxylamino) hexaaza cryptand, X = CNHOH (see Fig. 7.7).

The overall reaction can be represented as

$$Co(C_{14}H_{30}N_6)(NO_2)_2^{3+} \xrightarrow{e^-} Co(C_{14}H_{30}N_6)(NO_2)_2^{2+}$$

$$\xrightarrow{8H^+ + 7e^-} Co(C_{14}H_{30}N_6)(NHOH)_2^{3+} + 2H_2O. \qquad (7.7)$$

The second stage of (Eq. (7.7)) the production of a cobalt(III) complex in a reducing medium was unexpected. The step must involve reaction of the $\cdot SO_2^-$ entity with the NO_2 group or a subsequently reduced species. This step must precede the intramolecular oxidation of cobalt(II), which is the monitored process. Presumably the reactive intermediate formed scavenges, an electron from Co(II) with the same facility as from $\cdot SO_2^-$. It was also suggested that the relative ease of reduction of Co(dinosar)$^{3+}$ may result from the $-NO_2$ group channeling electrons to the metal center.

The kinetic data on the reduction of nitro-substituted cage complexes by dithionite [603] have been reinterpreted by Lay and Sargeson [604]. In particular, no evidence has been found for nitro groups mediating the rate of electron transfer to the Co(III) center.

7.3   Reduction of Halogen Compounds

Reactions of thioureas with oxyhalogen compounds have been attracted the attention of chemists for more than 100 years. Early studies have been performed mainly with preparative goals. In particular, it was shown that action of iodate and bromate on the thioureas is accompanied by their desulfurization [605]. The renaissance has begun about 30 years ago when Alamgir and Epstein [606] discovered that the oxidation of thiourea [606] and some of its derivatives (phenylthiourea, tetramethylthiourea) [607] by chlorite in acidic aqueous solution exhibit exotic kinetics. In a closed reaction vessel, for example, after a time delay, chlorine dioxide suddenly appears and its concentration shows several extrema as a function of time, if chlorite is used in excess over thiourea. Simoyi *et al.* [608] observed an traveling reaction front in an unstirred reaction mixture of thiourea and chlorite in acidic solution. The kinetics of the reaction between chlorite and thiourea in acidic medium has been studied by Epstein *et al.* in 1992 [609]. A 13-step mechanism (Eqs. (7.8)–(7.20)) was proposed with rate-determining step (Eq. (7.10)).

$$H^+ + ClO_2^- \rightleftharpoons HClO_2, \tag{7.8}$$

$$H^+ + (NH_2)_2CS \rightleftharpoons H-SC(NH_2)_2^+, \tag{7.9}$$

$$H^+ + (NH_2)_2CS + ClO_2^- \longrightarrow HOSC(NH)(NH_2) + HOCl, \tag{7.10}$$

$$H-SC(NH_2)_2^+ + ClO_2^- \longrightarrow HOSC(NH)(NH_2) + HOCl, \tag{7.11}$$

$$Cl_2 + H_2O \rightleftharpoons HOCl + Cl^- + H^+, \tag{7.12}$$

$$Cl_2 + 2ClO_2^- \longrightarrow 2ClO_2 + 2Cl^-, \tag{7.13}$$

$$HOCl + 2ClO_2^- + H^+ \longrightarrow 2ClO_2 + Cl^- + H_2O, \tag{7.14}$$

$$2ClO_2 + HOSC(NH)(NH_2) + H_2O$$
$$\longrightarrow HO_3SC(NH)(NH_2) + HOCl + Cl^- + H^+, \tag{7.15}$$

$$ClO_2^- + HOSC(NH)(NH_2) \longrightarrow HO_3SC(NH)(NH_2) + Cl^-, \tag{7.16}$$

$$ClO_2^- + HO_3SC(NH)(NH_2) + H_2O$$
$$\longrightarrow SO_4^{2-} + OC(NH_2)_2 + HOCl + H^+, \tag{7.17}$$

$$HOCl + HO_3SC(NH)(NH_2) + H_2O$$
$$\longrightarrow SO_4^{2-} + OC(NH_2)_2 + Cl^- + 3H^+, \tag{7.18}$$

$$HOSC(NH)(NH_2) + SC(NH_2)_2 \longrightarrow >CS-SC< + H_2O, \qquad (7.19)$$

$$>CS-SC< + ClO_2^- + H_2O + H^+$$

$$\longrightarrow 2HOSC(NH)(NH_2) + HOCl. \qquad (7.20)$$

Computer simulation using this mechanism gives a good agreement with experiment. As can be seen from the sequence of reactions (Eqs. (7.8)–(7.20)), the proposed mechanism does not assume the formation of TDO (aminoiminomethanesulfinic acid).

A year later Rábai *et al.* published the results of their study of reaction between thiourea and chlorine dioxide in acidic solution [610]. A 10-step mechanism incorporating a slow one-electron transfer from thiourea to ClO_2 to generate the $\cdot(NH)(NH_2)CS$ radical and subsequently more rapid reactions have been constructed and implemented in a computer simulation which provides a reasonably accurate fit to the observed kinetics curves:

$$(NH_2)_2CS + ClO_2 \longrightarrow \cdot(NH_2)(NH)CS + ClO_2^- + H^+, \qquad (7.21)$$

$$\cdot(NH_2)(NH)CS + ClO_2 + H_2O$$

$$\longrightarrow (NH_2)(NH)CSOH + ClO_2^- + H^+, \qquad (7.22)$$

$$2\cdot(NH_2)(NH)CS \longrightarrow (NH_2)(NH)CSSC(NH)(NH_2), \qquad (7.23)$$

$$HClO_2 \rightleftharpoons ClO_2^- + H^+, \qquad (7.24)$$

$$(NH_2)_2CS + HClO_2 \longrightarrow (NH)(NH_2)CSOH + HOCl, \qquad (7.25)$$

$$(NH_2)_2CS + HOCl \longrightarrow (NH)(NH_2)CSOH + Cl^- + H^+, \qquad (7.26)$$

$$(NH)(NH_2)CSOH + ClO_2 \longrightarrow \cdot(NH)(NH_2)CSO + HClO_2, \qquad (7.27)$$

$$\cdot(NH)(NH_2)CSO + (NH_2)(NH)CS$$

$$\longrightarrow (NH)(NH_2)CSOH + \cdot(NH_2)(NH)CS, \qquad (7.28)$$

$$(NH_2)_2CS + HOSC(NH)(NH_2)$$

$$\longrightarrow (NH)_2(NH)CSSC(NH)(NH_2) + H_2O, \qquad (7.29)$$

$$2\cdot(NH)(NH_2)CSO + H_2O$$

$$\longrightarrow (NH_2)(NH)CSOH + (NH)_2(NH)CSO_2H. \qquad (7.30)$$

The key species in this mechanism, formamidine sulfenic acid $((NH)(NH_2)CSOH)$, is formed in Eq. (7.22). Authors did not include in their mechanism the reaction of formamidine sulfinic (TDO) and formamidine sulfonic (thiourea trioxide) acids, though they noted that at lower pH and excess of ClO_2, thiourea is oxidized to formamidine sulfinic acid; in highly acidic condition and molar ratios of ClO_2 to thiourea of 5:1 and higher, some oxidation to formamidine sulfonic acid also occurs. The other important distinction of this mechanism from the mechanism of Epstein *et al.* [609] is the involvement of radicals.

In the same year, a general scheme which may be regarded as the common core in the mechanisms of the oscillatory chlorite ion–substrate reaction systems was suggested by Rábai and Orbán [611]. In the six-step model, the reactions of the substrate with ClO_2^-, HOCl and Cl_2O_2 and known reactions between the oxychlorine species were involved:

$$ClO_2^- + R + H^+ \longrightarrow HOCl + RO, \qquad (7.31)$$

$$HOCl + ClO_2^- + H^+ \longrightarrow Cl_2O_2 + H_2O, \qquad (7.32)$$

$$Cl_2O_2 + R + H_2O \longrightarrow 2HOCl + RO, \qquad (7.33)$$

$$R + HOCl \longrightarrow RO + H^+ + Cl^-, \qquad (7.34)$$

$$Cl_2O_2 + ClO_2^- \longrightarrow 2ClO_2 + Cl^-, \qquad (7.35)$$

$$Cl_2O_2 + H_2O \longrightarrow ClO_3^- + Cl^- + H^+. \qquad (7.36)$$

The general mechanism is capable of explaining the most exotic phenomena encountered in the chlorite–substrate reactions, such as the autocatalysis in batch and the simple and complex oscillations and chaos under flow conditions. However, Lengyel *et al.* suggested that the model requires more flexibility in describing the variable stoichiometry of the sulfur-containing reductants and that it may as well need to account for the reactivity of chlorine dioxide [612]. Model of Rábai and Orbán did not account for the possible pH dynamics from the oxidation of S(–II) species, including thiourea, and also failed to describe that birhythmicity appeared in the chlorite–thiourea reaction. The general model proposed by Thompson and coworkers predicted the existence of pH oscillations in the chlorite–thiourea reaction [228]. Their general model consists of three separate stages: negative hydrogen ion feedback process (S(–II) to S(0)), a transition of S(0) to S(IV) and positive proton feedback from S(IV) to S(VI). The observation of pH oscillations in this study provides experimental support to the sulfur (–II)-based reaction mechanisms suggested in the previous studies.

Gao and coworkers have shown that the reaction between chlorite and thiourea could display batch oligooscillation and CSTR oscillation of pH [613,614]. Batch pH peak has the same character with pH oscillation in a CSTR. The oxidation of thiourea produced intermediates such as

HOSC(NH)(NH$_2$), HO$_2$SC(NH)(NH$_2$), HO$_3$S(NH)(NH$_2$) and bisulfite. The valence change of sulfur has close relation with pH dynamics. It was also shown that a general model of sulfur(–II) oxidation [228], could simulate batch oligooscillations and CSTR oscillations.

One of the main drawbacks of all models mentioned before is that many values of rate constants used in them are not determined directly from the experiments and sometimes are very confusing from chemical point of view. To extend experimental basis for models of chlorite–thiourea reaction, Simoyi *et al.* have studied reaction of ClO$_2^-$ with TDO (formamidinesulfinic acid) at pH = 1–3 [615]. It was found that the formation of ClO$_2$ shows some oligooscillatory behavior in which, even in excess TDO, there is transient formation of ClO$_2$. The reaction is autocatalytic with respect to HOCl via the asymmetric Cl$_2$O$_2$ intermediate. The dynamics of the reaction is explained via a mechanism which is derived from the one used for the chlorite–thiourea reaction with intermediate formation of thiourea trioxide. But, again, this model involved rate constants whose values are very doubtful and have no experimental basis, especially for reactions on thiourea trioxide. Thus, rate constants of reactions of TTO with HOCl and ClO$_2^-$ have the same values (6.5×10^3 M^{-1}s^{-1}), which is very unlikely since HOCl is a much more potent oxidant than chlorite. Indeed, it was shown later that the reaction between chlorite and TTO is very slow [205]. Other drawbacks of the mentioned models has also been noted by Doona and Stanbury [616]. In particular, they have paid attention on the transition metal ions (e.g. Cu^{2+}) catalysis of thiourea reactions with chlorite, chlorine dioxide, iodate and bromate in batch conditions and emphasized that excluding the catalytic impact of adventitious metal impurities is necessary to determine accurate values of rate constants and to develop realistic chemical reaction mechanisms as well as to extrapolate these mechanistic studies to models simulating oscillatory phenomena displayed by these systems.

In the subsequent papers on kinetics of reactions between chlorite and substituted thioureas (phenyl- [617], trimethyl- [618], acetyl- [619] and tetramethyl- [105]) Simoyi and coworkers always used test for adventitious metal ion catalysis. However, not much difference was observed in the general reaction kinetics observed with distilled and deionized water. Reactions run with metal ion chelators such as ethylenediaminetetraacetic acid (EDTA) and deferroxamine did not offer different kinetics and dynamics. Surprisingly, the sets of reactions included in the models of reactions between substituted (phenyl-, trimethyl- and acetyl-) thioureas and chlorite consisted of many more reactions (31, 28 and 31, respectively) than in the case of unsubstituted thiourea (13 or 11, Refs. [609] and [620], respectively). Indeed, contrary to models published later [617–619], the earlier model [609] did not include reactions of TDO at all, what seems to be incorrect. But models suggested in the papers [617–619] could not also be considered as realistic since they contain too many unreliable and even contradictory data. Thus, rate constants for reactions of sulfoxylate with HOCl in the papers [617] and [618] have different values, but in case of acetylthiourea [619] reactions of sulfoxylate were not included at all. Data on reaction between tetramethylthiourea and chlorite seem to be more reliable and important [105]. Contrary to the other similar papers [617–619], a work of Chigwada *et al.* [105] contains direct experimental data on the sulfur-containing intermediates. Thus, ESI-MS data showed almost quantitative formation of the monoxide and negligible formation of the dioxide and trioxide of tetramethylthiourea. Unfortunately, having these data, authors did not offer the model of reaction "because activity of chlorine dioxide in acidic conditions could not be determined" [105].

In the attempts to fill a gap in the data on reactivity of chlorine dioxide, Gao *et al.* have studied its reaction with TDO [621,622]. In the course of these studies, indirect experimental evidence of the existence of a sluggish tautomerism of TDO in acidic aqueous solution was discovered. TDO slowly rearranges into aminoiminomethanesulfinic acid, but the reactivity of these two species towards chlorine dioxide marginally differs. It was clearly demonstrated that TDO reacts with chlorine dioxide in a relatively slow reaction. In contrast, reaction of aminoiminomethanesulfinic acid with chlorine dioxide in orders of magnitude is faster [621]. A 16-step kinetic model involving hypochlorous acid, chlorite ions and hydrogen sulfite ions as key intermediates that provide an autocatalytic cycle is proposed to account for the overall kinetic behavior observed, including the slow rearrangement of TDO [622].

Simoyi *et al.* have investigated the reaction between chlorite and hydroxymethanesulfinate (HMS) in weakly acidic and neutral solutions [623]. The reaction is very fast and is characterized by a short induction period (about 1 second) which is followed by a rapid and autocatalytic ClO_2 production. The rate-determining step is a 2-electron oxidation of HMS by ClO_2 to give the hydroxymethanesulfonate $HOCH_2SO_3^-$. A 19-step mechanism was used to simulate the reaction.

Kinetics of reaction between thiourea with iodate is also very interesting. The first detailed kinetic study of this reaction has been performed by Rábai and Beck [624]. It was shown that in the oxidation of thiourea by iodate in weakly acidic solution the concentration of iodide may exhibit several extrema. The rate-determining step is formation of thiourea monoxide and iodide:

$$IO_3^- + 3(NH_2)_2CS \longrightarrow I^- + 3(NH_2)_2CSO. \qquad (7.37)$$

Iodine is then produced in the reaction between iodate and iodide (Dushman reaction):

$$IO_3^- + 5I^- + 6H^+ \longrightarrow 3I_2 + 3H_2O. \qquad (7.38)$$

The kinetics of the iodine–thiourea reaction has also been studied and the reaction was found to be initiated by

$$(NH_2)_2CS + I_2 \longrightarrow (NH_2)(NH)CSI + H^+ + I^-, \qquad (7.39)$$

followed by

$$(NH_2)(NH)CSI + (NH_2)_2CS$$
$$\longrightarrow (NH_2)(NH)CSSC(NH)(NH_2) + H^+ + I^-. \qquad (7.40)$$

Thiourea monoxide formed in reaction (7.37) is oxidized by the excess of iodine to sulfate, ammonium ions and carbon dioxide via TDO and trioxide.

Taking into account the independently determined rate constants for some subsystems, the concentration–time profiles of iodide and iodine can be calculated. There is a good agreement between the experimental and theoretical curves. However, not all the rate coefficients of the kinetic model have been determined with sufficient precision. As an example it should be mentioned that reaction of TDO with iodate was studied later by Mambo and Simoyi [204]. The dynamics of the reaction was explained by a combination of three reactions: the first is the oxidation of TDO by iodate to give iodide, the second one is the Dushman reaction, and the third is the reaction of iodine and TDO. The relative rates of these three reactions determine the dynamics of the reaction. Oxidation of TDO with I_2 and I_3^- was also investigated. Oxidation of TDO by I_2 and I_3^- was found to be inhibited by H^+ and iodide ion. The proposed mechanism of the iodate–TDO reaction includes altogether 11 steps. Among them a seemingly important system (thiourea trioxide–iodine) was not studied at that time. Later Makarov and coworkers have found, that the rate of this reaction depends on the age of solution of TTO due to a possible formation of reactive species — bisulfite [205]. Aging the TDO solution has no significant effect on the rate of reaction with iodine. The same result has been received by Ojo and Simoyi for reaction between N,N$'$-dimethylthiourea dioxide with iodine [151]. These data show that reactions of TDO and TTO with iodine have different mechanisms. In case of thiourea trioxide iodine does not react directly with TTO itself, but with a product of its decomposition, bisulfite. However, TDOs react with iodine directly but not via decomposition products. Possible tautomerization to OH-form does not influence kinetics of their reactions with iodine. But, of course, a definite answer about the mechanism can only be given by experiments performed to determine the intermediates and end products of these reactions.

Thiourea–iodate reaction is a subsystem of thiourea–iodate–sulfite (TuIS) flow system which exhibits a rich variety of complex oscillations in pH [625–628]. In accordance with the experimental observations a simple kinetic model was suggested to account for these behaviors [628]. Previous model of Rábai *et al.* [629] was only capable of explaining simple oscillations in the CSTR. Based on some analogies of the behavior of thiosulfate and thiourea in chemical reactions Horváth *et al.* have suggested that the complexity of thiourea–iodate reaction serves as the main basis of the appearance of the complex oscillations in the TuIS system [628]. Authors proposed the following reaction mechanism:

$$H^+ + SO_3^{2-} \rightleftharpoons HSO_3^-, \tag{7.41}$$

$$3HSO_3^- + IO_3^- \longrightarrow 3SO_4^{2-} + I^- + 3H^+, \tag{7.42}$$

$$6Tu + IO_3^- + 6H^+ \longrightarrow 3Tu_2^{2+} + I^- + 3H_2O, \tag{7.43}$$

$$Tu + IO_3^- + 2H_2O \longrightarrow 2NH_4^+ + CO_2 + SO_3^{2-} + I^-, \tag{7.44}$$

where Tu_2^{2+} represents formamidine disulfide. Authors emphasized that model presented in the paper [628] is a simplification of the complex kinetics of the overall system. To set up a more accurate model the reaction kinetics between iodate and thiourea needs to be thoroughly investigated in detail.

Simoyi *et al.* have studied the kinetics of reaction between iodate and hydroxymethanesulfinate in acidic solutions [630]. The reaction presents clock reaction characteristics; iodide ions have a strong catalytic effect on the rate of the reaction by reducing the duration of the induction period. The direct reaction of aqueous iodine with HMS is very fast and produces sulfate, iodide and formaldehyde. Unfortunately, the proposed overall mechanism of HMS/iodate reaction is very confusing since it assumes intermediate formation of hydroxymethanesulfonate $HOCH_2SO_3^-$ in the HMS–iodine reaction (reaction of $HOCH_2SO_3^-$ with iodine is very slow in acidic solutions). Indeed, despite the using of HMS–iodine reaction for quantitative determination of hydroxymethanesulfinate for many years, the mechanism of this reaction is still unknown, as in the case of TDO–iodine reaction.

The reaction between periodate IO_4^- and thiourea has been studied in an unbuffered medium by Du *et al.* [631]. A variety of interesting phenomena including variable stoichiometries, characteristic "clock reaction" and oligooscillations or single peak oscillations in pH, the concentrations of both iodide and iodine were found in a closed reactor. When studied in an open reactor, the above reaction system displays fascinating damped oscillations and bistability in an unbuffered aqueous solution. A 12-step simplified mechanism has also been proposed here. The mechanism includes the consecutive formation of thiourea monoxide and TDO, but not thiourea trioxide.

In conclusion, let us consider the reactions of thioureas, TDO or HMS with bromate. Bromate-driven oscillators are the most thoroughly studied of all known homogeneous oscillating chemical systems [632]. Among the earliest known chemical oscillators is the Belousov–Zhabotinskii reaction in which bromate oxidizes an organic substrate in the presence of an acid and a metal ion catalyst [633,634]. The reaction between bromate and thiourea in acidic medium has been studied in a closed system and in the CSTR by Simoyi [635]. In closed systems, the reaction is characterized by an induction period. The induction period is followed by production of molecular bromine. In the CSTR the reaction displays sustained simple periodic oscillations in the bromine concentrations and the redox potential. A mechanism of reaction has not been proposed in this work. More detailed study of this reaction has been performed later by Simoyi, Epstein and Kustin [636]. A 14-step mechanism was proposed and used to simulate the observed kinetics. The rate-determining step for bromine appearance is formation of HOBr from the BrO_3^-–Br^- reaction. The oxidation of thiourea proceeds via oxygen additions on sulfur, successively forming thiourea mon-, di-, and trioxides and sulfate. But authors arbitrarily included it in the mechanism reaction between mon- and trioxides to form dioxide, which has no experimental basis. This reaction has been omitted from the bromate–methylthiourea [637] and bromate–trimethylthiourea [109] mechanisms, as new experimental data show that sulfur–sulfur interactions are insignificant in conditions of excess oxidant. As in case of chlorite [105], reaction of bromate with tetramethylthiourea occupies a special place [106], since there was no evidence for the formation of the tetramethylthiourea di- and trioxides in the oxidation pathway, but the formation of tetramethylthiourea monoxide was evidenced by the electrospray ionization mass spectrum of the dynamic reaction solution.

Simoyi's group have also studied bromate reactions with thiourea and N,N′-dimethylthiourea dioxides, as well as N,N′-dimethylthiourea trioxide [638,639]. In the earlier paper [638] authors proposed an oxidation pathway via thiourea trioxide without any experimental data on the reactivity of trioxide. Comparative study for N,N′-dimethylthiourea di- and trioxide showed, however, that trioxide is very inert in the oxidation even by bromine, and fast oxidation of dioxide can be explained by its decomposition to yield sulfoxylate. Indeed, conclusion about bypassing of trioxide in the oxidation of N,N′-dimethylthiourea and its dioxide is, seemingly, right, but data on the participation of sulfoxylate in dioxide oxidation are insufficient.

Jonnalagadda *et al.* have studied the reaction of hydroxymethanesulfinate with bromate and proposed 15-step mechanism [640]. Unfortunately, this mechanism contains many arbitrary reactions which have no experimental support. This especially regards reactions of hydroxymethanesulfonate.

Last problem we consider here, is a possibility of radical mechanisms of thiourea (thiourea oxides)–oxyhalogen reactions. In the papers mentioned before (see, for example, Ref. [638]), a radical mechanism was ruled out, i.e. oxidation of thioureas is considered as a successive oxygen atom transfer. But it is well known, that bromate–TDO system is used for the initiation of polymerization processes [641,642]. In these works formation of thiourea dioxide radicals is assumed. These data show that additional considerations of the mechanisms of thiourea (TDOs)–oxyhalogens is needed.

Chapter 8

Industrial Applications

8.1 Textile Industry

Sodium dithionite (hydrosulfite) belongs to one of the most important chemicals used in the textile industry. It is primarily applied as a reducing agent for the reduction of vat and sulfur dyes [643–652]. Dithionite reduces dyes to leuco forms, which are soluble in water and have a great affinity to the textile fiber. Vat dyes have mainly anthraquinone or indigoid/thioindigoid structures. Reduction of indigo dye is shown in Fig. 8.1 [652]. It is used with all vat dyes at temperatures ranging up from 30°C [650]. The advantage of sodium dithionite is that it causes fast reduction of vat dyes and it enables very short fixing times in various dyeing methods and produces levelness in continuous dyeings [651].

However, the disadvantage of sodium dithionite is that it cannot be recycled from waste waters [653]. It is unstable and very easily oxidized by atmospheric oxygen [1]. The result is that large amounts of dithionite are needed over the stoichiometric requirements of the reduction process although there have been studies on whether the amount of dithionite could be lowered and used more efficiently [648,649]. The generation of thiosulfate from dithionite has a corrosive effect on the waste lines [654]. The final product of dithionite oxidation — sulfate can cause damage to unprotected concrete pipes [648]. To improve the biocompatibility of vatting process, various electrochemical [653,655,656] and biological [657] reducing methods have been described as well as alternative chemical reductants such as iron(II) complexes [658], α-hydroxyketones [655] or borohydride [648]. But no reducing agent available today that can replace sodium dithionite in all areas. It is only used for special purposes (pad-steam processes and bath dyeings above 100°C) that have combinations with

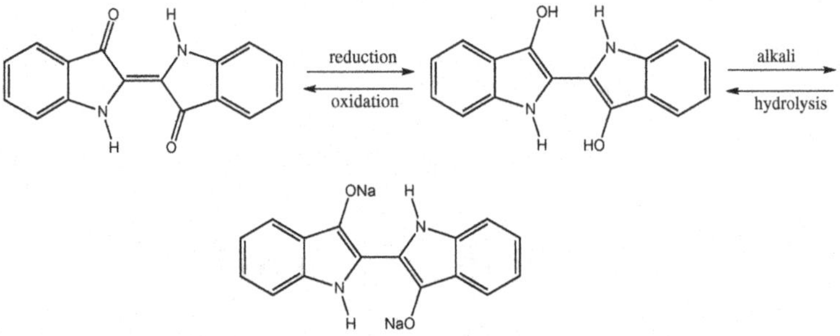

Fig. 8.1   Reaction of indigo dye to leucoform (soluble form)

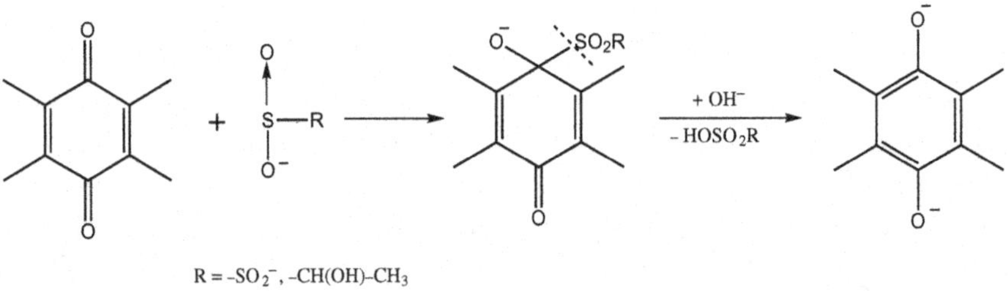

R = -SO$_2^-$, -CH(OH)-CH$_3$

Fig. 8.2 Reaction of anthraquinone derivatives with dithionite or α-hydroxyethanesulfinate

α-hydroxyalkanesulfinates — which are more stable to air oxygen — are also used. For textile printing, preference is given to sodium hydroxymethane-sulfinate (rongalite) [648].

The first detailed kinetic study of reaction between water-soluble anthraquinone derivative with sodium dithionite and α-hydroxyethane-sulfinate has been performed by Baumgarte [645]. The reaction is first order with respect to the dyestuff and the reducing agent. This would suggest that the rate-determining step is the addition of dithionite or α-hydroxyethanesulfinate to a keto-carbon atom of the dyestuff is followed by a fast release of the substituent added previously. This is followed by its conversion into the leuco compound. This scheme is illustrated in Fig. 8.2.

Baumgarte noted that if reactive species is sulfur dioxide anion radi-cal •SO$_2^-$ (as in many reactions of dithionite [1,659–661]), reaction should have been half order with respect to dithionite. But he performed his study at low temperature (8°C), much lower than other researchers (25–60°C) [659–661]. It is known, however, that reaction (8.1) strongly depends on

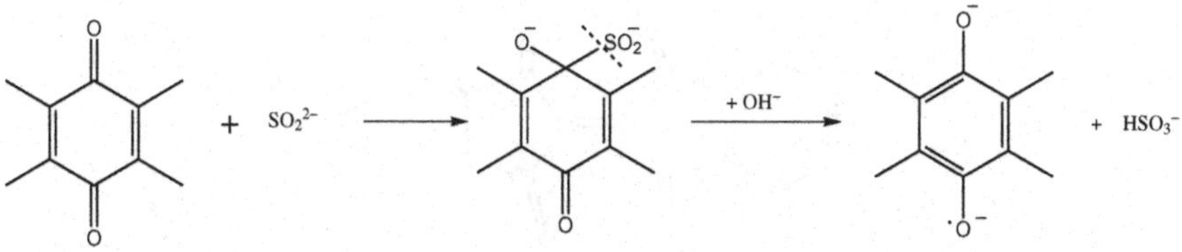

Fig. 8.3   Reaction of anthraquinone derivatives with sulfoxylate

the temperature [237] and concentration of anion radical increases with increasing temperature. Therefore, it is not surprising that at higher temperatures the impact of dithionite pathway is expected to be larger,

$$S_2O_4^{2-} \rightleftharpoons \ \cdot SO_2^-. \tag{8.1}$$

Similar to reductions with dithionite, mechanisms of oxidations of α-hydroxyalkanesulfinates strongly depend on temperature: the lower the temperature, the higher is the impact of α-hydroxyalkanesulfinate pathway. Conversely, at high temperatures, the main pathway is reduction by sulfoxylate SO_2^{2-} (SO_2H^-) (Fig. 8.3):

$$HOCHRSO_2^- \rightleftharpoons RCOH + SO_2H^-. \tag{8.2}$$

Srividya and coworkers have studied reduction of water-soluble derivative of indigo — indigocarmine [662]. This reaction takes place in two steps and direct involvement of $S_2O_4^{2-}$ (but not $\cdot SO_2^-$) is observed in both the places (Fig. 8.4). Authors have come to this conclusion because plots of observed rate constant vs. concentration of dithionite showed a linear dependence, with zero intercept, at all pH values studied. Unfortunately, we could not find any mention of temperature in these experiments and only can suspect that it was room temperature.

An important factor when dyeing with vat dyes is the rate of this process. In addition to temperature, pH, dispersion and crystal form of the dye particles, the rate of reduction of the dye can be influenced by accelerators, e.g. the anthraquinone type or its derivatives [648]. The reduction with dithionite proceeds sufficiently rapidly and such substances are not necessary for enhancing the rate. However, sodium hydroxymethanesulfinate reduces dyes at a much slower rate, and accelerators (catalysts) can be useful. Satisfactory results were obtained with cobalt dimethylglyoxime complexes [648]. Unfortunately, any sort of drawbacks came with a high cost.

It should be noted that though sodium hydroxymethanesulfinate is the most commonly used discharging agent in the textile industry, there is an

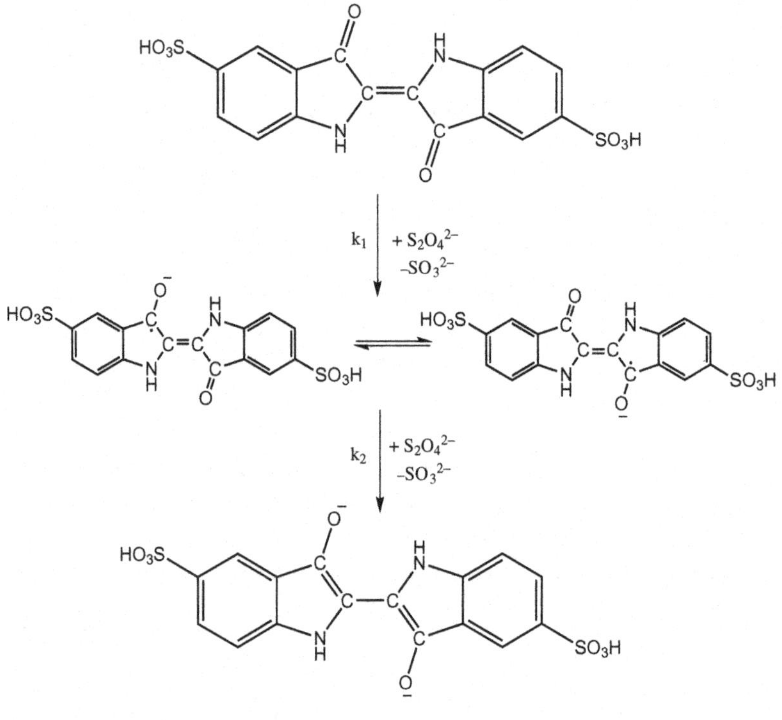

Fig. 8.4   Reaction of dithionite with indigocarmine

urgent need to develop an ecological alternative since HMS easily releases human carcinogen formaldehyde [78,79]. One of the best candidates is thiourea dioxide [78,97,114,663–666]. The advantages of using of thiourea dioxide are [1,78,663]: (1) TDO is a more ecofriendly chemical than hydroxymethanesulfinate and more stable in the solid state than dithionite; (2) while replacing dithionite with TDO alkali requirements can be reduced by approximately 20%; (3) at alkaline conditions TDO is a stronger reductant than dithionite and hydroxymethanesulfinate; (4) aqueous solutions of TDO are much more stable than solutions of dithionite, especially in presence of air; (5) unlike dithionite and hydroxymethanesulfinate no unpleasant odor problems are experienced during usage of TDO. Arifoglu and Marmer have developed a combined oxidative/reductive bleaching and dyeing process for wool in a single bath as an improved alternative to conventional bleaching and dyeing processes [97]. The newly developed single-bath process begins with an oxidative hydrogen peroxide bleaching followed by an addition of thiourea to the residual H_2O_2 in the same bath, i.e. synthesis of

TDO proceeds on site. Sulfoxylate, which is formed from TDO, affects the reductive bleaching. At the end of the reductive bleaching stage, a small amount of H_2O_2 is added to oxidize reductive sulfur species. This technology results in much brighter pastel shades upon dyeing due to more effective initial bleaching.

Recently Shao and coworkers have used TDO ground by the wet grinding process in presence of dispersants in discharging printing with decamethylcyclopentasiloxane (D5) as a solvent [78]. They have shown that as a substitute for water, D5 not only affords a kind of dispersion medium in grinding but also serves as an extremely advantageous stabilizer for TDO in practical applications. In D5 dispersion medium, the ground TDO could retain stable discharge ability for 100 days storage, which was significantly better than in H_2O. The fabrics discharged by ground TDO in D5 medium not only obtained good whiteness but also are easy to handle and have acceptable tensile strength loss. These results show that TDO has great perspectives of application in textile industry.

8.2 Paper Industry

Paper industry is one of the oldest and largest fields of application of sulfur-containing reducing agents. Their usage in this industry is described in detail in many books and reviews [52,667–675] (their number exceeds the number of books and reviews in rest of the fields altogether), therefore, here we consider this theme very briefly. Sodium dithionite (hydrosulfite) and thiourea dioxide are used mainly for bleaching and deinking of waste paper. Bleaching describes the destruction of chromophores. Therefore, any chemical reaction decreasing the conjugation of electrons in a colored molecule is a potential bleaching agent. Conjugation in a molecule can be altered by reduction and oxidation [667]. A perfect bleaching agent leaves the fiber unaffected and only takes care of chromophore destruction. Therefore, today, the list of compounds actually applied in bleaching of pulp is rather short. Potential reducing agents are almost totally represented by sulfur-containing compounds: sodium dithionite, thiourea dioxide and sulfur dioxide. The dominant compound used in bleaching processes nowadays is sodium dithionite.

Dithionite was used for the first time for bleaching mechanical pulps in the 1930s [52]. In an early study, Hosoya and coworkers assumed [676] that brightening of lignin with dithionite is based on the addition of bisulfite

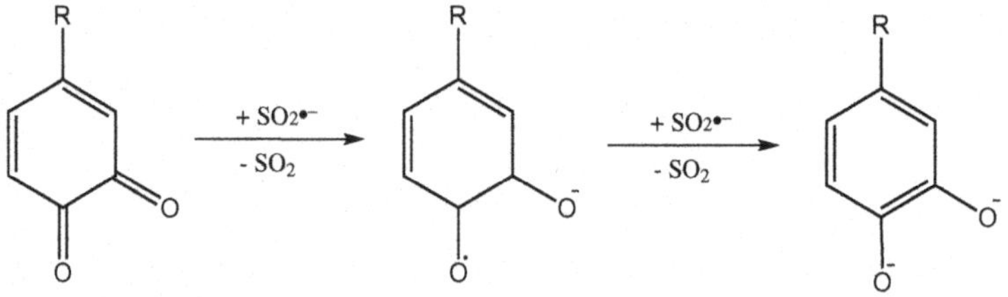

Fig. 8.5   Reaction of dithionite with lignin chromophores

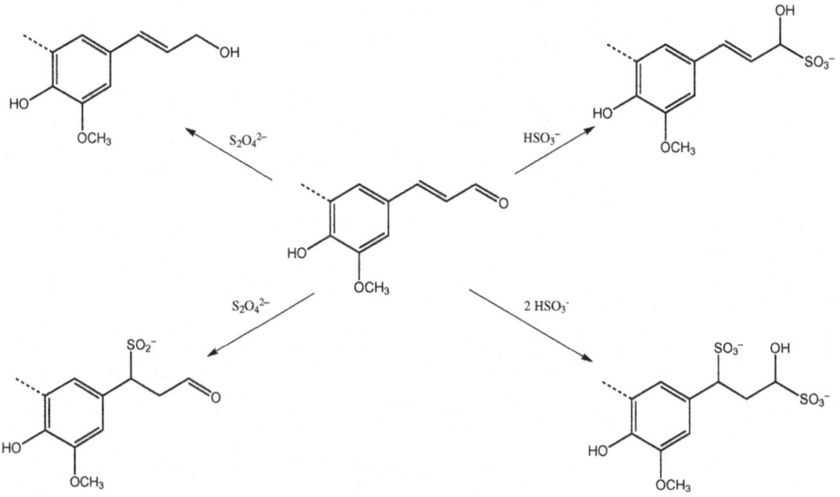

Fig. 8.6 Reduction of coniferaldehyde-type structures during dithionite bleaching

(which is generated from dithionite) to the conjugated double bonds and the reduction of chromophoric quinoid structures of lignosulfonates. But this assumption did not explain why dithionite is a much more effective bleaching agent than bisulfite. Later Pemberton and coworkers [677] have shown that the active reducing species in the reaction of dithionite and quinones is sulfur dioxide anion radical $\cdot SO_2^-$. Dence presented various reaction mechanisms during dithionite bleaching of ortho-benzoquinonoid structure (Fig. 8.5) and unsaturated coniferaldehyde-type structures (Fig. 8.6). Dithionite also reduces aldehyde and keto groups of lignin to alcohols [670] (reduction of aldehydes and ketones have already been discussed in a separate chapter of this book).

Bleaching with dithionite typically is done at moderately acidic pH, between pH 5 and 6, at temperature 60–80°C [667]. The addition of very high amounts might be detrimental due to the decomposition of dithionite with formation of very corrosive thiosulfate [667,678]. The other problem is oxygen. It is recommended to operate a dithionite stage (and the preparation of the dithionite solution) with closed systems to exclude air oxidation effects.

Hu and coworkers found that mechanical pulps prepared from beetle-infested, blue-stained lodgepole pine had a poor dithionite bleach response and observed a raise in the brightness ceiling on the bleached pulp [679]. The blue-stained thermomechanical pulp (TMP), when treated with low doses of peroxide, also exhibited a poor bleach response relative to the control. A method to overcome the brightness of dithionite-bleached TMP made from 25% blue-stained chips involves the addition of 0.2% of sodium borohydride to dithionite bleaching of the pulp [680]. A high bleaching end pH ($\approx$10.0) in peroxide bleaching removes most of the blue stains in TMP made from the blue-stained chips. Under optimal peroxide bleaching conditions, thermo- or chemithermomechanical pulp (CTMP) made from 50% or 100% blue-stained chips was bleached to the same brightness as TMP or CTMP from the green chips.

Hu, James and coworkers have discovered that water-soluble, tertiary hydroxyalkylphosphines or quaternary hydroxymethylphosphonium salts are effective bleaching and brightness stabilizing agents for mechanical pulps [674,679–682]. One of these agents, tetrakis-(hydroxymethyl) phosphonium sulfate (THPS), $[P(HOCH_2)_4]_2SO_4$, has been successfully tested on a commercial scale and used as a complementary bleaching agent to dithionite in the bleaching of a spruce stone-ground wood pulp to higher brightness. Use of THPS, $NaBH_4$ [683], or thiourea dioxide [684] as a complementary bleaching agent to dithionite may enable bleaching of mechanical pulps prepared from blue-stained chips to higher brightness.

Some authors, however, deny the usefulness of THPS for bleaching processes since results of Hu's group are difficult to repeat, the reaction products give the pulp an unpleasant smell, and the salt is expensive [667], and, most importantly, reaction of $(HOCH_2)_3P$ (and its associated precursor, $(HOCH_2)_4P^+Cl^-$) with $Na_2S_2O_4$ results in the formation of secondary $(HOCH_2)_2PH$ and primary $HOCH_2PH_2$ phosphines which are toxic [681]. Therefore, a wide-spread application is not likely to appear [667].

In the continuation of the work of Blechschmidt's group [684], Daneault and coworkers have shown that thiourea dioxide can be efficiently used as a bleaching chemical for softwood TMP [685]. The highest brightness was obtained at pH = 10, and the lowest yellowness at pH = 12. TDO-treated pulps were stable toward light-induced yellowing. Earlier they studied bleaching of TMP of spruce-balsam with TDO [686]. The results were compared to conventional bleaching with sodium dithionite; brightness obtained was approximately the same. The same group have also shown that in alkaline solution TDO significantly decolorizes yellow dye used in a local telephone directory [687].

TDO has found application in the bleaching and/or color stripping of recycled fibers [688] including mixed office waste [689]. Imamoglu has studied the role of printing color type and coating on the color stripping efficiency [690]. Thiourea dioxide and hydrogen peroxide were successfully used as reductive and oxidative bleaching agents, respectively. Standard offset printings were conducting on paper samples using cyan, magenta, yellow and black colors, and the effects of the deinking efficiency on the removal of each color were analyzed. Cyan color was found to be the most difficult one among the four main color printing inks stripping out in deinking process.

The usage of chemithermomechanical pulps is sometimes restrained by photoinduced discoloration that occurs when such pulps are exposed to daylight [691]. The α-carbonyl entities have been proposed as the main sensitizers of the photoyellowing process. Castellan and coworkers have studied the mechanism of action of some reducing agents, including sodium and zinc hydroxymethanesulfinates, thiourea dioxide, thioglycerol, 1-butylthioacetic acid, ascorbic acid, sodium hypophosphite, sodium metabisulfite, polyethylene glycol, in solution on compound 1 (see Fig. 8.7) which was chosen to mimic the behavior of bleached CTMP [691]. The discoloration of bleached CTMP, with the same additives, was studied for comparison. The results showed photochemical destruction of the colored substances by reducing agents, and sodium hydroxymethanesulfinate and ascorbic acid were the most effective ones. It was concluded that the reducing agents destroy peroxidic and colored entities which contribute to the discoloration process.

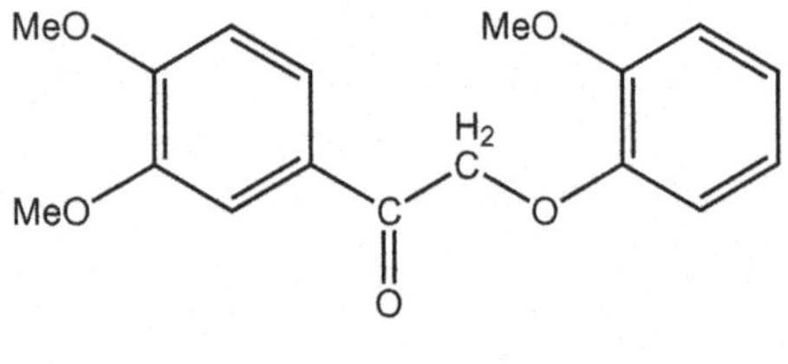

Fig. 8.7   Structure of compound 1

Note, in the conclusion, that according to Suess [667], sodium dithionite together with hydrogen peroxide is supposed to play the main role in the mechanical pulp bleaching in future. There is no alternative compound in sight to replace them. Thus, use of dithionite for the dyestuff destruction in wastepaper recycling is indispensable.

8.3 Polymerization Processes

Production of synthetic polymers is one of the main and well-known fields of application of sulfur-containing reducing agents. Initiation of emulsion polymerization at low temperature is usually based upon the production of free radicals during the course of reaction between ferrous iron and an organic hydroperoxide. The rate of this intrinsically very rapid reaction may be kept at a suitable level (thus preventing exhaustion of either ferrous iron or of hydroperoxide) by supplying ferrous iron continuously by reduction of ferric iron [692]. A considerable advance in the polymerization process was made by the discovery that water-soluble iron complex of ethylenediaminetetraacetic acid (Edta) or its salts is a powerful activator of peroxide-catalyzed emulsion polymerization [693]. The subsequent study confirmed the effectiveness of iron complex of Edta [694]. Because of the high degree of hydrolysis of trivalent iron, the chelation efficiencies of these chelants for ferric iron are low at the pH of polymerization. Thus, the major effect of the chelant must be the control of the ferrous ion level. Among inorganic reducing agents sodium hydroxymethanesulfinate (HMS, rongalite) [694] and sodium dithionite [692] have been found suitable for the reduction of ferric ethylenediaminetetraacetate (versenate, $FeEdta^{3+}$). The use of HMS as a reductant with $FeEdta^{3+}$ and diisopropylbenzene hydroperoxide (DIBH) was introduced by Howland *et al.* for Government Rubber Styrene (GRS) polymerization (copolymer of butadiene and styrene) [695]. HMS was found to give rapid rates of polymerization with both rosin and fatty acid soap emulsifications. Kolthoff and Meehan have shown that the dithionite reacts with $FeEdta^{3+}$ much faster than HMS (see Eqs. 8.3 and 8.4), but rate of polymerization in both cases have the same order of magnitude due to induced decomposition of both dithionite and hydroperoxide [692].

$$FeEdta^- + S_2O_4^{2-} \longrightarrow FeEdta^{2-} + \cdot S_2O_4^-, \qquad (8.3)$$

$$FeEdta^- + \cdot S_2O_4^- + 4OH^- \longrightarrow FeEdta^{2-} + 2SO_3^{2-} + 2H_2O. \qquad (8.4)$$

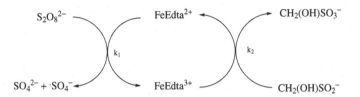

Fig. 8.8   Reactions in the system peroxodisulfate, ferrous sulfate, EDTA and HMS

Kerber and coworkers [696,697] studied reactions between HMS and hydrogen peroxide and in the system HMS, hydrogen peroxide and Fe^{3+}. Since they have shown that system hydrogen peroxide–HMS do not initiate polymerization, it has been concluded that reaction between them does not involve any radicals. It was shown that radicals responsible for initiation of polymerization are produced only when at least traces of heavy metals are available. Later this conclusion was confirmed by another research group [698]. Minato and Iwakawa have determined kinetic parameters of the reactions between HMS and Fe^{3+} (see Fig. 8.8) and between HMS and $FeEdta^{3+}$ in acidic and alkaline media [699]. The second-order rate constant for the former one at pH = 4 was found to be 76.0 $M^{-1}s^{-1}$, and for the latter one 0.00294 $M^{-1}s^{-1}$ (pH = 4) and 0.447 $M^{-1}s^{-1}$ (pH = 9) at 18°C. The reactions of dithionite with Fe^{3+} and $FeEdta^{3+}$ were very rapid, and their rates were only roughly estimated.

Preparation of acrylonitrile–butadiene–styrene (ABS) latexes using different hydroperoxide initiators together with HMS and iron Edta complex was studied by Daniels and coworkers [700]. There were no significant differences in polymerization kinetics or grafting efficiency between the t-butyl and cumene hydroperoxides (CHPs).

Kinetic study of redox reactions in the system containing ammonium peroxodisulfate, ferrous sulfate, EDTA and HMS is used in the copolymerization of tetrafluoroethylene and propylene in emulsion at low temperatures (25°C) has been performed by Kojima and Hisasue [701]. This study suggested that the working mechanism of the catalyst conforms to the scheme in Fig. 8.8, where the reduction of ferric ion seems to be rate determining step (i.e. $k_1 > k_2$).

The reactions of sodium hydroxymethanesulfinate with various monomers, including acrylates, methacrylates, styrene and vinyl acetate, were studied by Wang *et al.* [702]. It was shown that acrylates produced diacrylate sulfones in neutral conditions. The other monomers or reaction conditions did not result in a dimer.

Interfacial redox initiator system CHP/ferrous ion/EDTA/HMS, which could produce hydrophobic radicals and decrease homogeneous nucleation in the aqueous phase, was used to prepare mini-polymerization of styrene [703]. Polystyrene particles of narrow size distribution were obtained, and their molecular weights were also very narrow (polydispersity index $\approx$1.5–2.0). When higher CHP and HMS concentrations or higher reaction temperature were used, the polymerization rate and conversion became higher. The same system played a role of initiator in the graft copolymerization of 50/50 (w/w%) styrene/methyl methacrylate mixtures onto natural rubber seed latex [704]. The graft reaction mainly occurred by removal of hydrogen from the natural rubber followed by addition of macroradical units to that site. It was also confirmed that the graft copolymerization is a surface-controlled process.

Different redox initiator systems for emulsion polymerization of acrylates were investigated by Kohut-Svelko and coworkers [705]. In these systems, besides HMS, ascorbic acid, tetramethyl ethylene diamine (TMEDA), and sodium metabisulfites, they used Bruggolite FF6 (disodium salt of 2-hydroxy-2-sulfinatoacetic acid or disodium α-hydroxyethanoicsulfinate, or disodium glyoxylate sulfoxylate, $Na_2[O_2CCH(OH)SO_2]$) [80] and Bruggolite FF7 [706] (disodium salt of 2-hydroxy-2-sulfonatoacetic acid, disodium α-hydroxyethanoicsulfonate, or disodium glyoxylate bisulfite, $Na_2[O_2CCH(OH)SO_3]$) as reductants. The redox initiator systems also included two types of oxidants, persulfates and peroxides. Batch experiments showed that for systems using persulfates, the ammonium persulfate (APS)/TMEDA system provided the lower induction period and higher conversion, whereas for the systems with hydroperoxides, tert-butyl hydroperoxide (TBHP)/Bruggolite FF7, TBHP/HMS and H_2O_2/Bruggolite FF7 were the best alternatives. Interestingly, sulfonate derivative Bruggolite FF7 showed better results than more active reductant Bruggolite FF6 (if one considers them by analogy with hydroxymethanesulfinate and hydroxymethanesulfonate).

The redox couple (TBHP)/Bruggolite FF7 was recently assessed as initiator for reversible addition-fragmentation chain transfer (RAFT) polymerization at 25°C in H_2O/dioxane [707]. This couple allowed to achieve high monomer conversion (>99% within 8 h). The polymerization of 4-acryloylmorpholine was used as model reaction.

Liu and coworkers have studied seeded emulsion polymerization of n-butyl acrylate (n-BA) initiated by a redox initiator system: cumene hydroperoxide/ferrous sulfate hexahydrate/ethylene diaminetetraacetic acid monosodium salt/HMS [708]. It was found that the initial HMS concentration played a key role in the redox system for controlling the polymerization rate and conversions, and no significant effect on conversion could be seen with the increment addition of HMS. In the same year the stable emulsion of core/shell latex with little coagulum was prepared at low temperature with potassium persulfate, HMS, and 2,2′-azobis(2-(2-imidazolin-2-yl)propane)dihydrochloride as composite initiators by staged emulsion polymerization [709].

A modified redox system intended for initiation of radical polymerization of unsaturated polyester resins was investigated by Radenkov and coworkers [710]. The main feature of this modification was the presence of an additional accelerator — sodium hydroxymethanesulfinate. HMS acts as a reductant in hydrophilic medium. The interaction between HMS and the hydrophobic ingredients of the reactive mixture occur at the boundary surfaces between the different phases. HMS contributes for speeding up the polymerization process as well as compensates at least partly the toxic dimethyl aniline at the expense of HMS as efficient coaccelerator and also eliminates the inhibiting effect of even extremely high dozes of hydroquinone.

Sodium dithionite has been studied in polymer chemistry practically at the same time as HMS. We have already mentioned in an early paper [692] on analysis of applicability of sodium dithionite in polymerization. Now we shall consider more recent papers. The most valuable contribution regarding application of sodium dithionite in various polymerization processes has been made by Percec and Popov's groups [583,711–724]. In the early studies they used dithionite in the first non-transition metal-catalyzed living radical polymerization (LRP) of vinyl chloride (VC) in water at 25–35°C [711]. This polymerization is initiated with iodoform and catalyzed by $Na_2S_2O_4$ (see Fig. 8.9). Sulfur dioxide anion radical mediates the initiation and reactivation steps via a single-electron-transfer (SET) mechanism. The exchange between dormant and active propagating species also includes the degenerative chain transfer to dormant species (DT). In addition, the SO_2 released from $\cdot SO_2^-$ during the SET process can add reversibly to poly(vinyl chloride) (PVC) radicals and provide additional transient dormant radicals.

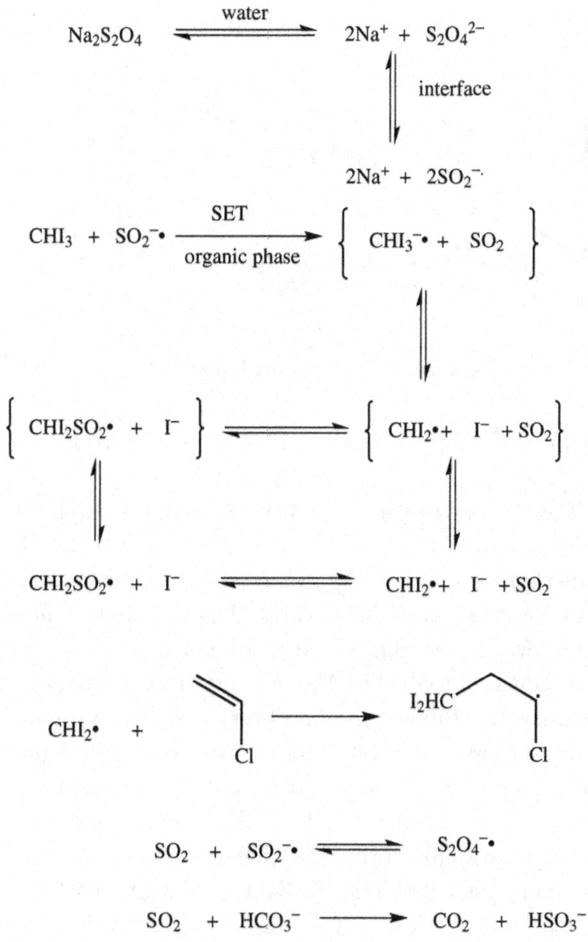

$$Na_2S_2O_4 \;\xrightleftharpoons{\text{water}}\; 2Na^+ + S_2O_4^{2-}$$

$$\Big\updownarrow \text{interface}$$

$$2Na^+ + 2SO_2^{-\cdot}$$

$$CHI_3 + SO_2^{-\cdot} \;\xrightarrow[\text{organic phase}]{\text{SET}}\; \left\{ CHI_3^{-\cdot} + SO_2 \right\}$$

$$\left\{ CHI_2SO_2^\cdot + I^- \right\} \;\xrightleftharpoons{}\; \left\{ CHI_2^\cdot + I^- + SO_2 \right\}$$

$$CHI_2SO_2^\cdot + I^- \;\xrightleftharpoons{}\; CHI_2^\cdot + I^- + SO_2$$

$$SO_2 + SO_2^{-\cdot} \;\xrightleftharpoons{}\; S_2O_4^{-\cdot}$$

$$SO_2 + HCO_3^- \;\longrightarrow\; CO_2 + HSO_3^-$$

Fig. 8.9   Initiation with CHI_3 and mediated by $Na_2S_2O_4$ SET reactions

In the following study, in addition to $Na_2S_2O_4$–$NaHCO_3$ catalytic system authors used electron transfer cocatalysts (ETC) 1,1-dialkyl-4,4-bipyridinum dihalides or alkyl viologens (AlkV^{2+}) [712]. The role of ETC is to accelerate the transfer of an electron from the H_2O phase to the organic phase and also to transfer the iodide anion from the organic phase to the aqueous phase (Fig. 8.10) (AlkV^{2+} are more soluble in water than AlkV$^{+\cdot}$ that are more soluble in organic solvents, consequently, viologens are used to transport electrons from an aqueous to an organic phase). This

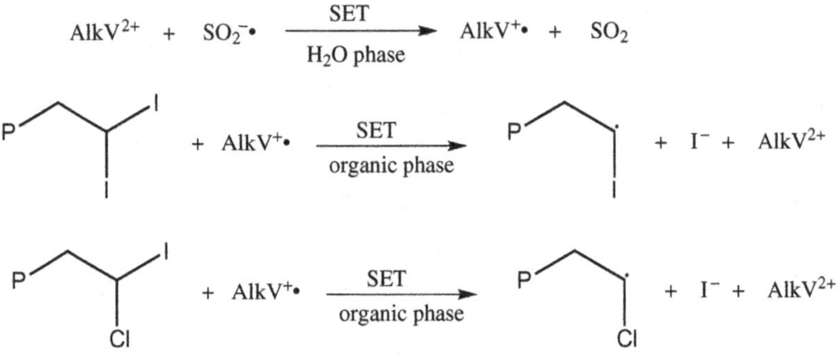

AlkV2+ + SO2-• SET H2O phase AlkV+• + SO2
P I I + AlkV+• SET organic phase P I + I- + AlkV2+
P I Cl + AlkV+• SET organic phase P Cl + I- + AlkV2+

Fig. 8.10 SET step mediated by ETC

accelerates the SET step of VC polymerization in H_2O/tetrahydrofuran (THF) at 25°C.

In 2005, Percec and coworkers continued studies of SET-degenerative chain transfer mediated living radical polymerization (DTLRP) of VC initiated with methylene iodide CH_2I_2 instead of iodoform and catalyzed by sodium dithionite in water at 35°C [713]. The reactivity and initiator efficiency of CH_2I_2 was lower than those of iodoform.

An acceleration of SET–DTLRP of VC in water can be accomplished at 43°C in presence of a catalytic amount of cetyltrimethylammonium bromide as a phase-transfer catalyst [714]. The accelerated method provides access to thermally stable PVC. However, the higher concentration of growing radicals reduces the functionality of the PVC chain ends and, therefore, limits its use in additional synthetic applications.

Besides polymerization of VC, $Na_2S_2O_4$–$NaHCO_3$ catalytic system acts as a SET agent and facilitates the SET–DTLRP of acrylates in water initiated with iodoform at room temperature [715]. Authors concluded that the resulting α,ω-di(iodo)polyacrylates can be used as macroinitiators for the synthesis of complex polyacrylate architectures. Indeed, later α,ω-di(iodo)poly(2-ethylhexyl acrylate) [α,ω-di(iodo)P2EHA] as a macroinitiator for the SET–DTLRP of VC catalyzed with $Na_2S_2O_4$ in H_2O. This synthetic method provides for the first time access to a diversity of block copolymers and other complex architectures based on combinations of thermoplastic PVC and elastomeric polyacrylates [716].

Dithionite-catalyzed polymerization has been used for preparation of poly(n-butyl acrylate) (PBA) in water [717]. The methodology studied in this work represents a possible route to prepare well-tailored macro-molecules made of butyl acrylate in an environmental friendly reaction medium. Moreover, such materials can be subsequently functionalized. Later Popov and coworkers have applied SET/DTLRP in water catalyzed by sodium dithionite for preparation of poly(lauryl acrylate) [718], poly(ethyl acrylate) [719], and poly(2-methoxyethyl acrylate) [720] as well as studied the influence of the isomer structures of butyl acrylate monomer on SET/DTLRP [721].

In the most recent papers of Popov's group sodium dithionite and supplemental activator with $Cu(II)Br_2/Me_6TREN$ catalytic system have been used in atom transfer radical polymerization (ATRP) of 2-(diisopropylamino)ethyl methacrylate (DPA) at 40°C in a mixture isopropanol-water [722,723]. Earlier they have shown that sodium dithionite is a more powerful reducing agent that can reduce Cu(II) species into Cu(I) more effectively than sodium metabisulfite and sodium bisulfite [583,724]. Well-controlled polymers required $Na_2S_2O_4$ to be slowly and continuously fed to the reaction mixture [723].

The other link between $Na_2S_2O_4$ and polymers is the usage of dithionite on exchange resins in redox reactions [725]. One of the several advantages given by ion exchange resins as supported reagents is the possibility of using water soluble inorganic salts as reagents in organic solvents. Reductions of C=C and C=O double bonds in several unsaturated substrates were thus performed with the dithionite anion on exchange resins, chemo- and regioselectivities were set as examples in a series of monoterpenic compounds.

Thiourea dioxide is less employed in polymerization processes than sodium dithionite and hydroxymethanesulfinate. The first examples of application of TDO are relevant to 1970s and 1980s. Hebeish and coworkers have studied hydrogen peroxide-thiourea dioxide redox system induced grafting of 2-methyl-5-vinylpyridine onto oxidized celluloses [726]. Excluding thiourea dioxide from the polymerization system offsets grafting onto cotton cellulose while considerable grafting takes place on the various oxidized celluloses and their further modified samples. Authors assumed that the presence of a reducing agent such as thiourea dioxide as a cocatalyst with H_2O_2 will accelerate decomposition of the latter species to produce free radicals due to one electron transfer with concomitant cleavage of the O–O bond under the attack of this nucleophile. The reaction involved may be presented as follows:

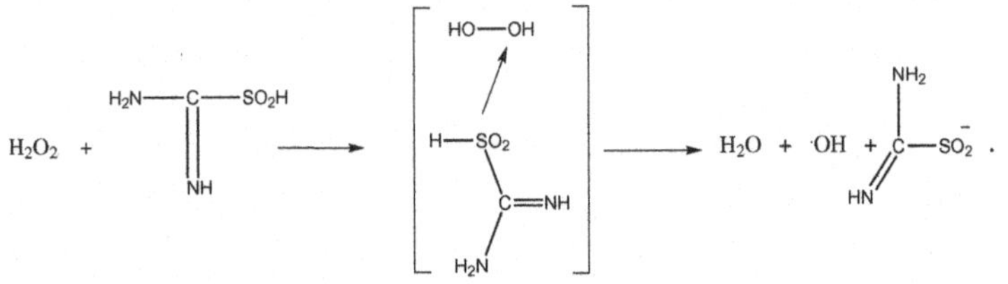

$$H_2O_2 + \underset{\underset{NH}{\overset{\|}{}}}{\overset{H_2N-\overset{|}{C}-SO_2H}{}} \longrightarrow \left[\underset{H_2N}{\overset{HO-OH}{\underset{C=NH}{H-SO_2}}} \right] \longrightarrow H_2O + {}^{\cdot}OH + \underset{HN}{\overset{NH_2}{C-SO_2^-}} . \qquad (8.5)$$

Polymerization onto cellulose can be initiated by the hydroxyl radical, or the thiourea dioxide radical, or both. Later Hebeish and coworkers have used Fe^{2+}–thiourea dioxide–H_2O_2 redox system for the polymerization of glycidyl methacrylate, methacrylic acid, acrylamide and their mixtures as well as mixtures of glycidyl methacrylate with acrylamide, acrylonitrile, butylmethacrylate, or styrene with cotton fabric [727,728], and various vinyl monomers with flax fibers [729]. The initiation mechanism involves the following reactions (cell–OH is cellulose) [729]:

$$Fe^{2+} + H_2O_2 \longrightarrow OH^- + {}^{\cdot}OH + Fe^{3+}, \qquad (8.6)$$

$$Fe^{3+} + (NH_2)(NH)CSO_2H \longrightarrow (NH_2)(NH)CSO_2{}^{\cdot} + Fe^{2+} + H^+, \qquad (8.7)$$

$${}^{\cdot}OH + (NH_2)(NH)CSO_2H \longrightarrow (NH_2)(NH)CSO_2{}^{\cdot} + H_2O, \qquad (8.8)$$

$${}^{\cdot}OH + cell-OH \longrightarrow cell-O{}^{\cdot} + H_2O, \qquad (8.9)$$

$$(NH_2)(NH)CSO_2{}^{\cdot} + cell-OH \longrightarrow cell-O{}^{\cdot} + (NH_2)(NH)CSO_2H. \qquad (8.10)$$

Grafting of methacrylic acid onto loomstate cotton fabric can be induced by potassium permanganate/thiourea dioxide redox system [730]. The other system, potassium bromate/TDO has been used for preparation of poly(acrylic acid) hydroxyethyl cellulose [poly(AA)–HEC] graft copolymer by polymerizing acrylic acid with hydroxyethyl cellulose [642] and poly(acrylic acid)/starch hydrogel [731]. This crosslinked hydrogel was used for the removal of the Cd^{2+} from its aqueous solution by adsorption.

By analogy of dithionite/alkyl viologen system [712], redox system thiourea dioxide/$NaHCO_3^-$/octyl viologen (OV^{2+}) was used as a catalyst in SET/DTLRP of VC [732]. In both cases the structure of polymers is identical. The differences between these two processes are a lower rate of polymerization, a lower monomer conversion, and a lower molecular weight of the final polymer obtained for the TDO/$NaHCO_3$/OV^{2+} catalytic system.

This means that the concentration of growing radicals is lower. Another dissimilarity is that $Na_2S_2O_4$ catalyzes SET–DTLRP of VC without the aid of ETC [711,712], whereas TDO does not. Indeed, the initial active species from the TDO is the sulfoxylate SO_2^{2-} (see Eqs. 8.11–8.15), but not sulfur dioxide anion radical as in the case of dithionite. To form the SET agent $\cdot SO_2^-$ from SO_2^{2-}, the presence of an oxidizing agent (for example, OV^{2+}) is required. The absence of this oxidizing agent in the water phase provides a possible explanation for the inactivity of TDO in SET–DTLRP of VC in the absence of ETC.

$$(NH_2)_2CSO_2 \rightleftharpoons (NH_2)(NH)CSO_2H, \tag{8.11}$$

$$(NH_2)(NH)CSO_2H + H_2O \xrightarrow{NaHCO_3} (NH_2)_2CO + SO_2^{2-} + 2H^+, \tag{8.12}$$

$$SO_2^{2-} + OV^{2+} \xrightarrow{SET,\ water} SO_2^- \cdot + OV^+ \cdot, \tag{8.13}$$

$$SO_2^- \cdot + OV^{2+} \xrightarrow{SET,\ water} SO_2 + OV^+ \cdot, \tag{8.14}$$

$$SO_2^{2-} + OV^+ \cdot \xrightarrow{SET,\ water} SO_2^- \cdot + OV^0. \tag{8.15}$$

Thiourea dioxide can be not only used as a part of initiation system, but also a part of polymer itself. Thus, polyorganylsilsesquioxanes containing carbofunctional fragments of TDO were prepared by oxidation of N,N$'$-bis[3-(triethoxysilyl)propyl]thiourea and N-acetyl-N$'$-[3-(triethoxysilyl)propyl]thiourea with a 40–50% solution of hydrogen peroxide [733]. Sorption of Ag(I), Au(III), Pd(II) and Pt(IV) on the polymers prepared was studied.

Chapter 9

Miscellaneous

In analytical chemistry, in addition to the analysis of nitrogen-containing compounds mentioned in a previous chapter, TDO can be employed for determination of organic disulfides [734], rhodium and iridium [735,736]. Thiourea dioxide can as well be used as an effective toning agent in photography [737].

Recently, TDO was suggested to be used for bitumen modification [46–48]. Thermo-gravimetric analysis demonstrated the formation of new chemical compounds, most probably originated through reactions between products from thiourea dioxide thermal decomposition and some highly polar bitumen molecules [47]. Viscosity was always seen to increase after the addition of TDO. Also, modification enhanced elastic properties and thermal susceptibility at high in-service temperatures and yielded improved binder resistance to thermal cracking by a decrease in its glass transition temperature [46]. It was shown that thiourea dioxide addition produces changes in the bitumen colloidal nature by means of hydrogen-bond network [48], which improve its flexibility at low in-service temperatures, and consequently its resistance to thermal cracking under loading. In fact, dynamic bending tests indicated a remarkable decrease in the value of binder glass transition temperature, which was further corroborated by differential scanning calorimetry [47]. These results suggest that the usage of thiourea dioxide may become a promising alternative to that of other chemical modifiers for the paving industry.

Strong reducing properties of sodium dithionite were exploited in a new, rather unexpected application in 2016. Mallick et al. suggested to use $Na_2S_2O_4$ in nano or micromotors [738]. Synthesis of autonomously moving nano and micromotors is an immediate challenge in nanoscience and nanotechnology. In their work, authors reported a system based on soft-oxometalates (SOMs) which is very easy to synthesize and moves autonomously in response to chemical stimuli like that of a reducing agent — dithionite. The redox active Mo(VI) sites of SOMs are used for oxidizing dithionite to generate SO_2 to propel the micromotors. In this reaction Mo(VI) oxidizes dithionite to SO_2 in accordance with the following reaction (shown schematically):

$$Mo_7^{VI}O_{24} + S_2O_4^{2-} \rightarrow \text{molybdenium blue} + 2SO_2. \qquad (9.1)$$

An interesting paper has been published recently by Gao *et al.* [739]. For the first time thiourea dioxide has been explored as an efficient chemiluminescence (CL) coreactant. The luminol/thiourea dioxide CL can be significantly enhanced by Co^{2+}. This system enables sensitive detection of thiourea dioxide, luminol, and Co^{2+}, particularly selective Co^{2+} detection. These results became unexpected because luminol manifests luminescent properties in the presence of oxidant, for example, peroxide, but not reductant as TDO. It is important to note that oxygen did not influence the CL of luminol. To explain their data, authors assumed that peroxide can be generated in the course of decomposition of TDO:

$$(NH_2)(NH)CS(O)OH \longrightarrow (NH_2)(NH)CS-OOH \longrightarrow (NH_2)_2CS + O_2. \qquad (9.2)$$

The increase in CL intensities with increasing pH from 10.3 to 11.9 was attributed to the faster generation of effective hydroperoxide intermediate from TDO at higher pH and the deprotonation of luminol. It should be noted, however, that there is no experimental evidence of formation of this intermediate, and additional studies have to be done.

The ability of TDO to produce oxygen (see Ref. [282]) and, possibly, peroxide compounds may also have other consequences. Recently several incidents have been reported involving the spontaneous exothermic decomposition of thiourea dioxide during the shipping of drums in containers [740,741]. Products of the reaction caused widespread contamination around the surrounding areas that required expensive remedial action. The probability of these hazardous consequences increases at high temperatures and high humidity. This shows that the detailed study of conditions assisting the evolution of oxygen or formation of peroxides from TDO is absolutely mandatory.

Note, however, that direct experimental evidence of formation of oxygen in aqueous solutions of TDO is not available yet. Therefore, mentioned hazardous consequences might be explained by formation of other gases.

Bibliography

1. S. V. Makarov, Recent Trends in the Chemistry of Sulfur-Containing Reducing Agents, *Russ. Chem. Rev.* **70**, pp. 885–895 (2001).
2. Y. Wang, L. Sun and B. Fugetsu, Thiourea Dioxide as a Green Reductant for the Mass Production of Solution-Based Graphene, *Bull. Chem. Soc. Jpn.* **85**, pp. 1339–1344 (2012).
3. C. K. Chua, A. Ambrosi and M. Pumera, Graphene Oxide Reduction by Standard Industrial Reducing Agent: Thiourea Dioxide, *J. Mater. Chem.* **22**, pp. 11054–11061 (2012).
4. C. K. Chua and M. Pumera, Chemical Reduction of Graphene Oxide: A Synthetic Chemistry Viewpoint, *Chem. Soc. Rev.* **43**, pp. 291–312 (2014).
5. N. Pan, J. Deng, D. Guan, Y. Jin and C. Xia, Adsorption Characteristics of Th(IV) Ions on Reduced Graphene Oxide from Aqueous Solutions, *Appl. Surf. Sci.* **287**, pp. 478–483 (2013).
6. N. Pan, D. Guan, Y. Yang, Z. Huang, R. Wang, Y. Jin and C. Xia, A Rapid Low-Temperature Synthetic Method Leading to Large-Scale Carboxyl Graphene, *Chem. Eng. J.* **236**, pp. 471–479 (2014).
7. Q. Ma, J. Song, C. Jin, Z. Li, J. Liu, S. Meng, J. Zhao and Y. Guo, A Rapid and Easy Approach for the Reduction of Graphene Oxide by Formamidinesulfinic Acid, *Carbon* **54**, pp. 36–41 (2013).
8. N. I. Park, W.-S. Park, S. B. Lee, S. M. Lee and D.-W. Chung, Comparative Studies on Three Kinds of Reductants Applicable for the Reduction of Graphene Oxide, *Appl. Chem. Eng.* **26**, pp. 99–103 (2015).
9. J. Wang, T. Zhou, H. Deng, F. Chen, K. Wang, Q. Zhang and Q. Fu, An Environmentally Friendly and Fast Approach to Prepare Reduced Graphite Oxide with Water and Organic Solvents Solubility, *Colloids Surf. B: Biointerfaces* **101**, pp. 171–176 (2013).
10. D. A. Davies, J. Silver, A. Vecht, P. J. Marsh and J. A. Rose, A Novel Method for the Synthesis of ZnS for Use in the Preparation of Phosphors for CRT Devices, *J. Electrochem. Soc.* **148**, pp. H143–H148 (2001).
11. Y. U. Jeong and A. Manthiram, Synthesis of Nickel Sulfides in Aqueous Solutions Using Sodium Dithionite, *Inorg. Chem.* **40**, pp. 73–77 (2001).
12. H. Wang, Q. Li and C. Gao, Preparation of Nanometer Nickel Powder from Spent Electroless Nickel Plating Baths by Using Thiourea Dioxide as a Green Reductant, *J. Clean. Prod.* **84**, pp. 701–706 (2014).

13. I. N. Terskaya, D. S. Salnikov, S. V. Makarov, O. V. Yarovaya and S. A. Lilin, Chemical Synthesis of Stable Nano-Sized Water-organic Copper Dispersions, *Prot. Met.* **44**, pp. 468–470 (2008).

14. S. Kotha and P. Khedkar, Rongalite: A Useful Green Reagent in Organic Synthesis, *Chem. Rev.* **112**, pp. 1650–1680 (2012).

15. O. Louis-Andre and G. Gelbard, Mise au Point Réductions Chimiques par le Dithionite de Sodium, *Bull. Soc. Chim. Fr.* **4**, pp. 565–577 (1986).

16. S. Kumar, S. Verma, S. L. Jain and B. Sain, Thiourea Dioxide (TUD): A Robust Organocatalyst for Oxidation of Sulfides to Sulfoxides with TBHP under Mild Reaction Conditions, *Tetrahedron Lett.* **52**, pp. 3393–3396 (2011).

17. S. Verma, S. L. Jain and B. Sain, PEG-Embedded Thiourea Dioxide (PEG.TUD) as a Novel Organocatalyst for the Highly Efficient Synthesis of 3,4-Dihydropyrimidinones, *Tetrahedron Lett.* **51**, pp. 6897–6900 (2010).

18. S. Verma and S. L. Jain, Thiourea Dioxide Catalyzed Multi-Component Coupling Reaction for the One Step Synthesis of Naphthopyran Derivatives, *Tetrahedron Lett.* **53**, pp. 6055–6058 (2012).

19. S. Verma, S. Kumar, S. L. Jain and B. Sain, Thiourea Dioxide Promoted Efficient Organocatalytic One-Pot Synthesis of a Library of Novel Heterocyclic Compounds, *Org. Biomol. Chem.* **9**, pp. 6943–6948 (2011).

20. S. Verma, R. Singh, D. Tripathi, P. Gupta, G. M. Bahuguna and S. L. Jain, Thiourea Dioxide with TBHP: A Fruitful and Greener Recipe for the Catalytic Oxidation of Alcohols, *RSC Adv.* **3**, pp. 4184–4188 (2013).

21. S. Kumar, S. L. Jain and B. Sain, Thiourea Promoted Cobalt-Catalyzed Hydrolysis of Imines: Dual Activation via Organocatalysis and Metal Catalysis, *RSC Adv.* **2**, pp. 789–791 (2012).

22. M. Ghanshang, S. S. Mansoor and K. Aswin, Thiourea Dioxide: An Efficient and Reusable Organocatalyst for the Rapid One-Spot Synthesis of Pyrano[4,3-b]pyran Derivatives in Water, *Chin. J. Catal.* **35**, pp. 127–133 (2014).

23. A.-C. Schmidt, F. W. Heinemann, C. E. Kefalidis, L. Maron, P. W. Roesky and K. Meyer, Activation of SO_2 and CO_2 by Trivalent Uranium Leading to Sulfite/Dithionite and Carbonate/Oxalate Complexes, *Chem. Eur. J.* **20**, pp. 13501–13506 (2014).

24. S. V. Klementyeva, M. T. Gamer, A.-C. Schmidt, K. Meyer, S. N. Konchenko and P. W. Roesky, Activation of SO_2 with $[(\eta^5\text{-}C_5Me_5)_2Ln(THF)_2]$ (Ln=Eu, Yb) Leading to Dithionite and Sulfinate Complexes, *Chem. Eur. J.* **20**, pp. 13497–13500 (2014).

25. S. V. Klementyeva, N. Arleth, K. Meyer, S. N. Konchenko and P. W. Roesky, Dithionite and Sulfinate Complexes from the Reaction of SO_2 with Decamethylsamarocene, *New J. Chem.* **39**, pp. 7589–7594 (2015).

26. H. Nakai, M. Mizuno, T. Nishioka, N. Koga, K. Shiomi, Y. Miyano, M. Irie, B. K. Breedlove, I. Kinoshita, Y. Hayashi, Y. Ozawa, T. Yonezawa, K. Toriumi and K. Isobe, Direct Observation of Photochromic Dynamics in the

Crystalline State of an Organorhodium Dithionite Complex, *Angew. Chem. Int. Ed.* **45**, pp. 6473–6476 (2006).

27. H. Nakai, T. Nonaka, Y. Miyano, M. Mizuno, Y. Ozawa, K. Toriumi, N. Koga, T. Nishioka, M. Irie and K. Isobe, Photochromism of an Organorhodium Dithionite Complex in the Crystalline-State: Molecular Motion of Pentamethylcyclopentadienyl Ligands Coupled to Atom Rearrangement in a Dithionite Ligand, *J. Am. Chem. Soc.* **130**, pp. 17836–17845 (2008).

28. H. Nakai, M. Hatake, Y. Miyano and K. Isobe, The Absolute Asymmetric Photoisomerization of a Photochromic Dithionite Complex in Chiral Crystals, *Chem. Comm.* **45**, pp. 2685–2687 (2009).

29. H. Nakai, S. Uemura, Y. Miyano, M. Mizuno, M. Irie and K. Isobe, Photoreactivity of Crystals of a Rhodium Dithionite Complex with Ethyltetramethylcyclopentadienyl Ligands: Crystal Surface Morphology Changes and Degradation, *Dalton Trans.* **40**, pp. 2177–2179 (2011).

30. H. Nakai, M. Hatake and K. Isobe, Crystalline-State Photochromism of a Dithionite Complex in Chiral Crystal, *Acta Cryst. Sect. A* **A64**, p. C424 (2008).

31. P. Bruna, A. Decken, S. Greer, F. Grein, H. D. B. Jenkins, B. Mueller, J. Passmore, T. A. Paulose, J. M. Rautiainen, S. Richardson and M. J. Schriver, Synthesis of $(TDAE)(O_2SSO_2)(s)$ and Discovery of $(TDAE)(O_2SSSSO_2)(s)$ Containing the First Polythionite, $[O_2SSSSO_2]^{2-}$, *Inorg. Chem.* **52**, pp. 13651–13662 (2013).

32. A. Vegas, J. F. Liebman and H. D. B. Jenkins, Unique Thermodynamic Relationships for $\Delta_f H^0$ and $\Delta_f G^0$ for Crystalline Inorganic Salts. I. Predicting the Possible Existence and Synthesis of Na_2SO_2 and Na_2SeO_2, *Acta Cryst. Sect. B* **B68**, pp. 511–527 (2012).

33. C. F. Schönbein, Ueber Farbenveräderungen, *J. Prakt. Chem.* **61**, pp. 193–224 (1854).

34. P. Schützenberger, Sur un Nouvel Acide de Soufre, *Compt. Rendus* **69**, pp. 196–201 (1869).

35. P. Schützenberger, Sur un Nouvel Acide de Soufre, *Ann. Chim. Phys.* **20**, pp. 351–361 (1870a).

36. R. Schubart, *Sulfinic Acids and Derivatives in Ullmann's Encyclopedia of Industrial Chemistry*. Wiley-VCH, Weinheim (2012).

37. E. de Barry Barnett, VII.- The Action of Hydrogen Dioxide on Thiocarbamides, *J. Chem. Soc.* **97**, pp. 63–65 (1910).

38. T. Frangsmyr, *Nobel Lectures, Chemistry 1971–1980*. World Scientific Publishing Co., Singapore (1993).

39. A. E. Miller, J. J. Bischoff and K. Pae, Chemistry of Aminoiminomethanesulfinic and -Sulfonic Acids Related to the Toxicity of Thioureas, *Chem. Res. Toxicol.* **1**, pp. 169–174 (1988).

40. P. Messinger and H. Greve, Synthese Symmetrischer Sulfone aus Natrium-Hydroxymethanesulfinat und Mannichbasen, *Synthesis*, **1977**, pp. 259–261 (1977).

41. J. S. Sessler, P. A. Gale and W.-S. Cho, *Anion Receptor Chemistry*, 1st edn. RSC Publishing, Cambridge (2006).
42. O. V. Serdyuk, C. M. Heckel and S. B. Tsogoeva, Bifunctional Primary Amine-thioureas in Asymmetric Organocatalysis, *Org. Biomol. Chem.* **11**, pp. 7051–7071 (2013).
43. S. V. Makarov and R. Silaghi-Dumitrescu, Sodium Dithionite and Its Relatives: Past and Present, *J. Sulfur Chem.* **34**, pp. 444–449 (2013).
44. S. V. Makarov, A. S. Makarova and R. Silaghi-Dumitrescu, *Sulfoxylic and Thiosulfurous Acids and Their Dialkoxy Derivatives in Patai's Chemistry of Functional Groups, Peroxides*, Vol. 3, 1st edn. Wiley, Hoboken (2014).
45. S. V. Makarov, A. K. Horváth, R. Silaghi-Dumitrescu and Q. Gao, Recent Developments in the Chemistry of Thiourea Oxides, *Chem. Eur. J.* **34**, pp. 14164–14176 (2014).
46. A. A. Cuadri, P. Partal, F. J. Navarro, M. Garcia-Morales and C. Gallegos, Influence of Processing Temperature on the Modification Route and Rheological Properties of Thiourea Dioxide-Modified Bitumen, *Energ. Fuel.* **25**, pp. 4055–4062 (2011).
47. A. A. Cuadri, P. Partal, F. J. Navarro, M. Garcia-Morales and C. Gallegos, Bitumen Chemical Modification by Thiourea Dioxide, *Fuel* **90**, pp. 2294–2300 (2011).
48. A. A. Cuadri, V. Carrera, M. A. Izquierdo, M. Garcia-Morales and F. J. Navarro, Bitumen Modifiers for Reduced Temperature Asphalts: A Comparative Analysis between Three Polymeric and Non-Polymeric Additives, *Constr. Build. Mater.* **51**, pp. 82–88 (2014).
49. T. L. Davies, Paul Schutzenberger, *J. Chem. Educ.* **6**, pp. 1403–1414 (1929).
50. J. Wisniak, Paul Schützenberger, *Educ. Quim.* **26**, pp. 57–65 (2015).
51. J. G. de Vries and R. M. Kellogg, Reduction of Aldehydes and Ketones by Sodium Dithionite, *J. Org. Chem.* **45**, pp. 4126–4129 (1980).
52. P. Bajpai, *Environmentally Benign Approaches for Pulp Bleaching.* Elsevier Science, Amsterdam (2012).
53. Hydrosulfite — the evergreen among textile auxiliaries turns a hundred. ChemEurope.com, 2004. http://www.chemeurope.com/en/news/36309/hydrosulfite-the-evergreen-among-textile-auxiliaries-turns-a-hundred.html.
54. The Organisation for Economic Cooperation and Development (OECD). Screening Information Data Set (SIDS), Sodium Dithionite, http://www.inchem.org/documents/sids/sids/7775146.pdf.
55. M. Ding and L. Feng, Determination of Organic Anions in Waste Liquid of Sodium Hydrosulfite Production by Ion Chromatography, *J. Chromatogr. A* **839**, pp. 233–237 (1999).
56. J. Shang, Y. Zhang, F. Zhou, F. Lv, F. Han, J. Lu, X. Meng, P. K. Chu, Z. Ye and J. Xing, Analysis of Hazardous Organic Residues from Sodium Hydrosulfite Industry and Utilization as Raw Materials in a Novel Solid Lubricant Production, *J. Hazard. Mater.* **198**, pp. 65–69 (2011).
57. M. Goliath and B. O. Lindgren, Mechanism of Reduction of Sulphur Dioxide by Formic Acid, *Acta Chem. Scand.* **16**, pp. 570–574 (1962).

58. C. E. McKenna, W. G. Gutheil and W. Song, A Method for Preparing Analytically Pure Sodium Dithionite. Dithionite Quality and Observed Nitrogenase-Specific Activities, *Biochim. Biophys. Acta* **1075**, pp. 109–117 (1991).

59. D. M. Oglesby, R. L. Ake and W. P. Kilroy, Chemical and Physical Properties of Lithium Dithionite–Cathode Discharge Product in the Li/SO_2 Cell in *Proceedings of the 30th Power Sources Symposium.* Atlantic City, NJ, pp. 127–130 (1982).

60. S. A. Chmielewski and D. W. Benneth, A New Synthon for Main Group Metal Dithionites in a Non-Aqueous Environment, *Synth. React. Inorg. Met.-Org. Chem.* **16**, pp. 95–103 (1986).

61. O. Brunck, Ueber die Einwirkung von Hydroschwefligsaurem Natrium auf Metallsalze, *Justus Liebigs Ann. Chem.* **336**, pp. 281–298 (1904).

62. A. Magnusson and L.-G. Johansson, The Crystal Structure of Tin(II) Dithionite, $Sn_2(S_2O_4)_2$, *Acta Chem. Scand. A* **36**, pp. 429–433 (1982).

63. S. M. Lough and J. W. McDonald, Synthesis of Tetraethylammonium Dithionite and Its Dissociation to the Sulfur Dioxide Radical Anion in Organic Solvents, *Inorg. Chem.* **26**, pp. 2024–2027 (1987).

64. T. Mincey and T. G. Taylor, A Novel Hemin Reducing System: Crown Ether Complexes of Sodium Dithionite and Their Uses as Reducing Agents in Nonaqueous Solvents, *Bioinorg. Chem.* **9**, pp. 409–420 (1978).

65. P. J. Bruna, S. Greer, J. Passmore and J. M. Rautiainen, Evidence for $[18–Crown–6Na]_2[S_2O_4]$ in Methanol and Dissociation to $Na_2S_2O_4$ and 18-Crown-6 in the Solid State; Accounting for the Scarcity of Simple Oxy Dianion Salts of Alkali Metal Crown Ethers in the Solid State, *Inorg. Chem.* **50**, pp. 1491–1497 (2011).

66. E. Potteau, E. Levillain and J. P. Lelieur, Electrochemical Study of Polythionite Solutions in HMPA, *J. Electroanal. Chem.* **436**, pp. 271–275 (1997).

67. H. Nakai, Y. Miyano, Y. Hayashi and K. Isobe, Synthesis and Structural Characterization of a Photochromic Dirhodium Dithionite Complex: $[(Cp^{Ph}Rh)_2(\mu-CH_2)_2(\mu-O_2SSO_2)]$ ($Cp^{Ph}=\eta^5-C_5Me_4Ph$), *Mol. Cryst. Liq. Cryst.* **456**, pp. 63–70 (2006).

68. Y. Miyano, H. Nakai, M. Mizuno and K. Isobe, Substitution Effects of Cp Ring Benzyl Groups on Photoisomerization of a Rhodium Dithionite Complex in the Crystalline State, *Chem. Lett.* **37**, pp. 826–827 (2008).

69. G. J. Kubas, H. J. Wasserman and R. R. Ryan, Reduction of Sulfur Dioxide by $C_5R_5M(CO)_3H$ (M = Mo, W; R = H, Me). Chemistry and Structures of $(C_5H_5)Mo(CO)_3(SO_2H)$, the First Example of Insertion of SO_2 into a Metal-Hydride Bond, and $[(C_5Me_5)Mo(CO)_3]_2(\mu-S_2O_4)$, a Sulfur-Bonded Dithionite Complex, *Organometallics* **4**, pp. 2012–2021 (1985).

70. K. Reinking, E. Dehnel and H. Labhardt, Zur Constitution der Aldehydschwefligsauren Salze und der Hydroschwefligen Säure, *Chem. Ber.* **38**, pp. 1069–1080 (1905).

71. M. Mulliez, C. Naudy and R. Wolf, Synthese d'α-Hydroxysulfinates, *Tetrahedron* **49**, pp. 2469–2476 (1993).

72. V. F. Postnikov and T. I. Kunin, Studies on Production of Rongalite, *Zh. Prikl. Khim.* **13**, pp. 185–190 (1940).

73. J. R. Nooi, P. C. van Hoeven and W. P. Haslinghuis, Reactions of Photoexcited SO$_2$; Preparation of α-Substituted Alkanesulphinic Acids, *Tetrahedron Lett.* **11**, pp. 2531–2534 (1970).

74. W. Küppers, Rongal A in Flash Ageing, *J. Soc. Dyers Colour.* **78**, pp. 597–603 (1962).

75. S. V. Makarov, B. Y. Sokolova and V. V. Budanov, Kinetic Parameters of Stability and Reactivity of Sodium Aminoalkanesulfinates, *Zh. Obshch. Khim.* **55**, pp. 724–730 (1985).

76. M. E. Mulliez and C. Naudy, Synthese de Sels d'α-Aminoacides Sulfiniques Substitues á lázote par un Groupe Méthane- ou p-Toluéne-Sulfonyle, *Tetrahedron* **50**, pp. 5401–5412 (1994).

77. F. J. M. Dujols and M. E. Mulliez, Simple Synthesis of Variously N-Protected α-Aminomethanesulfinate Salts, *J. Org. Chem.* **61**, pp. 5648–5649 (1996).

78. L. Zhou, J. Shan, X. Liu and J. Shao, Study of the Application of Modified Thiourea Dioxide Discharge Agent in D5 Non-Aqueous Medium, *Color. Technol.* **131**, pp. 149–156 (2015).

79. K. Karthikeyan and B. Dhurai, New Method of Discharge Printing on Cotton Fabrics Using Horseradish Peroxidase, *AUTEX Res. J.* **11**, pp. 61–65 (2011).

80. M. Lubik and P. Fithian, VOC-Free Alternative for Green Coatings, *Paint and Coatings Ind.* **28**, pp. 36–38 (2012).

81. V. K. Sharma, Potassium Ferrate(VI): An Environmentally Friendly Oxidant, *Adv. Environ. Res.* **6**, pp. 143–156 (2001).

82. S. Sahu, P. R. Sahoo, S. Patel and B. K. Mishra, Oxidation of Thiourea and Substituted Thioureas: A Review, *J. Sulfur Chem.* **32**, pp. 171–197 (2011).

83. J. Böeseken, Étude sur les Oxydes de Thiourée, I. Sur le Dioxyde de Thiourée, CS(NH$_2$)$_2$O$_2$, *Rec. Trav. Chim.* **55**, pp. 1040–1043 (1936).

84. J. Böeseken, Étude sur les Oxydes de Thiourée. Iv, *Rec. Trav. Chim.* **67**, pp. 603–621 (1948).

85. W. Walter and G. Randau, Über die Oxydationprodukte von Thiocarbonsäureamiden, XIX. Thioharnstoff-S-Monoxide, *Justus Liebigs Ann. Chem.* **722**, pp. 52–79 (1969).

86. W. Walter and G. Randau, Über die Oxydationprodukte von Thiocarbonsäureamiden, XIX. Thioharnstoff-S-Dioxide, *Justus Liebigs Ann. Chem.* **722**, pp. 80–97 (1969).

87. W. Walter and G. Randau, Über die Oxydationprodukte von Thiocarbonsäureamiden, XIX. Thioharnstoff-S-Trioxide (Guanylsulfonsäurebetaine), *Justus Liebigs Ann. Chem.* **722**, pp. 98–109 (1969).

88. W. Walter and C. Rohloff, Über die Struktur der Thioamide und Ihrer Derivate, XXXII Darstellung Sowie IR- und H-NMR-spektroskopische Untersuchung von 2-Aminobenzamiden und Thiobenzamiden, *Justus Liebigs Ann. Chem.* **1975**, pp. 295–304 (1975).

89. D. De Filippo, G. Ponticelli, E. F. Trogu and A. Lai, Spectrochemical Study of Aminoiminomethanesulphinic Acid and Related N,N′-Substituted Derivatives, *J. Chem. Soc. Perkin Trans. 2* **2**, pp. 1500–1502 (1972).

90. E. Y. Yarovenko and R. P. Lastovskii, Synthesis and Properties of Alkyl (aryl) Substituted Formamidine Sulfinic Acids, *Zh. Org. Khim.* **6**, pp. 947–949 (1970).

91. J. J. Havel and R. Q. Kluttz, A Synthesis of Fomamidinesulfinic Acids and Formamidines, *Synth. Commun.* **4**, pp. 389–393 (1974).

92. A. E. Miller and J. J. Bischoff, A Facile Conversion of Amino Acids to Guanidino Acids, *Synthesis* **1986**, pp. 777–779 (1986).

93. C. A. Maryanoff, R. C. Stanzione, J. N. Plampin and J. E. Mills, A Convenient Synthesis of Guanidines from Thioureas, *J. Org. Chem.* **51**, pp. 1882–1884 (1986a).

94. C. A. Maryanoff, R. C. Stanzione and Plampin, Reactions of Oxidized Thioureas with Amine Nucleophiles, *Phosphorus Sulfur* **27**, pp. 221–232 (1986b).

95. G. Gattow and W. Manz, Oxydationprodukte von Thioharnstoff, *Z. Anorg. Allg. Chem.* **561**, pp. 66–72 (1988).

96. K. Kim, Y. T. Lin and H. S. Mosher, Monosubstituted Guanidines from Primary Amines and Aminoiminomethanesulfonic Acid, *Tetrahedron Lett.* **29**, pp. 3183–3186 (1988).

97. M. Arifoglu and W. N. Marmer, Sequential Oxidative/Reductive Bleaching and Dyeing of Wool in a Single Bath at Low Temperature, *Text. Res. J.* **62**, pp. 123–130 (1992).

98. K. Rittstieg, W. Somitsch and K.-H. Robra, Quantification of the Bleaching Agent FAS and its Decomposition Products in Industrial Waste Water by HPLC in *Proceedings of 5th Symposium on Instrumental Analysis*. Pécs, Hungary, p. P7 (1999).

99. K. Rittstieg, K.-H. Robra and W. Somitsch, Aerobic Treatment of a Concentrated Urea Wastewater with Simultaneous Stripping of Ammonia, *Appl. Microbiol. Biotechnol.* **56**, pp. 820–825 (2001).

100. G. Zhang, S. Zheng and G. Lu, Heat Effect of Synthetic Process of Thiourea Dioxide from Thiourea and Hydrogen Peroxide, *Huadong Ligong Daxue Xuebao* **22**, pp. 206–210 (1996).

101. W. Li and T. Nonaka, Paired Electrosynthesis of Aminoiminomethanesulfonic Acids, *Electrochim. Acta* **44**, pp. 2605–2612 (1999).

102. Y. Hu, J. Feng, Y. Li, Y. Sun, L. Xu, Y. Zhao and Q. Gao, Kinetic Study on Hydrolysis and Oxidation of Formamidine Disulfide in Acidic Solutions, *Sci. China Chem.* **55**, pp. 235–241 (2012).

103. Q. Y. Gao, G. P. Wang, Y. Y. Sun and I. R. Epstein, Simultaneous Tracking of Sulfur Species in the Oxidation of Thiourea by Hydrogen Peroxide, *J. Phys. Chem. A* **112**, pp. 5771–5773 (2008).

104. L. L. Poulsen, R. M. Hyslop and D. M. Ziegler, S-Oxygenation of N-Substituted Thioureas Catalyzed by the Pig Liver Microsomal FAD-Containing Monooxygenase, *Arch. Biochem. Biophys.* **198**, pp. 78–88 (1979).

105. T. Chigwada, W. Mbiya, K. Chipiso and R. H. Simoyi, S-Oxygenation of Thiocarbamides V: Oxidation of Tetramethylthiourea by Chlorite in Slightly Acidic Media, *J. Phys. Chem. A* **118**, pp. 5903–5914 (2014).

106. R. O. Ajibola and R. H. Simoyi, S-Oxygenation of Thiocarbamides IV: Kinetics of Oxidation of Tetramethylthiourea by Aqueous Bromine and Acidic Bromate, *J. Phys. Chem. A* **115**, pp. 2735–2744 (2011).

107. D. Li, H. Li, Y. Luo, K. Li, Q. Meng, M. Armand and L. Chen, Non-Corrosive, Non-Absorbing Organic Redox Couple for Dye-Sensitized Solar Cells, *Adv. Funct. Mater.* **20**, pp. 3358–3365 (2010).

108. Y. Liu, J. R. Jennings, M. Parameswaran and Q. Wang, An Organic Redox Mediator for Dye-Sensitized Solar Cells with Near Unity Quantum Efficiency, *Energy Environ. Sci.* **4**, pp. 564–571 (2011).

109. T. R. Chigwada, E. Chikwana, T. Ruwona, O. Olagunju and R. H. Simoyi, S-Oxygenation of Thiocarbamides 3: Nonlinear Kinetics in the Oxidation of Trimethylthiourea by Acidic Bromate, *J. Phys. Chem. A* **111**, pp. 11552–11561 (2007).

110. W. Walter and K.-P. Rueβ, Über die Oxidationsprodukte von Thiocarbonsäureamiden, XXIX. Konfiguration Alkyl- und Aryl-Substituierter Thioharnstoff-S-Trioxide, *Liebigs Ann. Chem.* **1974**, pp. 243–252 (1974).

111. W. Walter and K.-P. Rueβ, Über die Oxidationsprodukte von Thiocarbonsäureamiden, XXX. Konfiguration und Behinderte Rotation in Aryl-Alkyl-Substitutierten Thioharnstoff-S-Trioxiden und Formamidiniumsalzen; Trennung der Geometrischen Isomeren Aryl-Trialkyl-Substituierter Thioharnstoff-S-Trioxide, *Liebigs Ann. Chem.* **1974**, pp. 253–273 (1974).

112. M. Hoffmann and J. O. Edwards, Kinetics and Mechanism of the Oxidation of Thiourea and N,N$'$-dialkylthioureas by Hydrogen Peroxide, *Inorg. Chem.* **16**, pp. 3333–3338 (1977).

113. S. K. Saha and D. J. Greenslade, Isothiocarbamido Radicals from Thiourea: Electron Spin Resonance Spectroscopy of N-Benzylidene-t-butylamine-N-Oxide and 5,5-Dimethyl-1-Pyrroline-N-Oxide Spin Adducts, *Bull. Chem. Soc. Jpn.* **65**, pp. 2720–2723 (1992).

114. M. Arifoglu, W. N. Marmer and R. L. Dudley, Reaction of Thiourea with Hydrogen Peroxide: ^{13}C NMR Studies of an Oxidative/Reductive Bleaching Process, *Text. Res. J.* **62**, pp. 94–100 (1992).

115. D. M. Ziegler, Intermediate Metabolites of Thiocarbamides, Thioureylenes and Thioamides: Mechanism of Formation and Reactivity, *Biochem. Soc. Trans.* **6**, pp. 94–96 (1978).

116. K. Ziegler-Skylakakis, S. Nill, J. F. Pan and U. Andrae, S-Oxygenation of Thiourea Results in the Formation of Genotoxic Products, *Environ. Mol. Mutagen.* **31**, pp. 362–373 (1998).

117. U. Andrae and H. Greim, *Initiation and Promotion in Thyroid Carcinogenesis in Tissue Specific Toxicity: Biochemical Mechanisms.* Academic Press, N.Y. (1992).

118. U. Andrae, *Reactive Intermediates of Xenobiotics in Thyroid. Formation and Biological Consequences.* Plenum Press, N.Y. (1996).

119. W. Zhou, K. Peng and F.-M. Tao, Theoretical Mechanism for the Oxidation of Thiourea by Hydrogen Peroxide in Gas State, *J. Mol. Struct.*: *THEOCHEM* **821**, pp. 116–124 (2007).

120. D. Chatterjee, S. Rothbart and R. van Eldik, Selective Oxidation of Thiourea with H_2O_2 Catalyzed by $[Ru^{III}(edta)(H_2O)]^-$: Kinetic and Mechahistic Studies, *Dalton Trans.* **42**, pp. 4725–4729 (2013).

121. P. Sarkar and D. Chatterjee, Oxidation of Thiourea by Peroxomonosulfate Ion Catalyzed by a Ruthenium (III) Complex, *Transit. Met. Chem.* **41**, pp. 9–13 (2016).

122. W. Wang, M. N. Schuchmann, H.-P. Schuchmann, W. Knolle, J. von Sonntag and C. von Sonntag, Radical Cations in the OH-Radical-Induced Oxidation of Thiourea and Tetramethylthiourea in Aqueous Solution, *J. Am. Chem. Soc.* **121**, pp. 238–245 (1999).

123. G. O. Schenck and H. Wirth, Photooxydation von Thioharnstoff zu Amino-Imino-Methan-Sulfinsäure, *Naturwissenschaften* **40**, p. 141 (1953).

124. I. Kraljic and H. E. A. Kramer, On the Reaction of Singlet Molecular Oxygen with Allylthiourea in Aqueous Solution, *Photochem. Photobiol.* **27**, pp. 9–12 (1978).

125. G. Crank and A. Mursyidi, Oxidations of Thioureas with Photochemically Generated Singlet Oxygen, *J. Photochem. Photobiol. A* **64**, pp. 263–271 (1992).

126. G. Crank and M. I. H. Makin, A New Method for Converting Thiourea and Monosubstituted Thioureas into Cyanamides: Desulphurisation by Superoxyde Ion, *J. Chem. Soc. Chem. Commun.* **20**, pp. 53–54 (1984).

127. Y. H. Kim and G. H. Yon, Novel Desulphurization of 1,3-Disubstituted Thioureas by Superoxide Anion ($\bullet O_2^-$): One-Step Synthesis of 1,2,3-Trisubstituted Guanidines from 1,3-Disubstituted Thioureas, *J. Chem. Soc. Chem. Commun.* **19**, pp. 715–716 (1983).

128. J. P. James, G. B. Quistad and J. E. Casida, Ethylenethiourea S-Oxidation Products: Preparation, Degradation, and Reaction with Proteins, *J. Agric. Food. Chem.* **43**, pp. 2530–2535 (1995).

129. J. D. Dunitz, The Structure of Sodium Dithionite and the Nature of the Dithionite Ion, *Acta Cryst.* **9**, pp. 579–586 (1956).

130. C. T. Kiers and A. Vos, The Nature of the S–S Bonds in Different Compounds. V. The Crystal Structure of ZnS_2O_4 Pyridine, *Acta Crystallogr. B* **34**, pp. 1499–1504 (1978).

131. L. Peter and B. Meyer, The Structure of Dithionite Ion, *J. Mol. Struct.* **95**, pp. 131–139 (1982).

132. H. Takahashi, N. Kaneko and K. Miwa, Raman and Infrared Studies of the Structure of the Dithionite Ion in Aqueous Solution and Force-Constants of $S_2O_x^{2-}$-type Ions, *Spectrochim. Acta A* **38**, pp. 1147–1153 (1982).

133. W. C. Hodgemann, J. B. Weinrach and D. W. Bennett, Spectroscopic Evidence for a Centrosymmetric Dithionite Anion in the Solid State: Vibrational Spectroscopy of Tetraethylammonium Dithionite, *Inorg. Chem.* **30**, pp. 1611–1614 (1991).

134. K. I. Carter, J. B. Weinrach and D. W. Bennett, A Density Functional Study of Sulfoxy Anions Containing Sulfur–Sulfur Bonds, *J. Am. Chem. Soc.* **115**, pp. 10981–10987 (1993).

135. J. B. Weinrach, D. R. Meyer, J. J. T. Guy, P. E. Michalski, K. L. Carter, D. S. Grubisha and D. W. Bennett, A Structural Study of Sodium Dithionite and Its Ephemeral Dihydrate: A New Conformation for the Dithionite Ion, *J. Cryst. Spectr. Res.* **22**, pp. 291–301 (1992).

136. S. A. Chmielewski, J. B. Weinrach, D. W. Bennett and W. P. Kilroy, Evidence for the Formation of a New Form of Lithium Dithionite under Rigorous Nonaqueous Conditions, *J. Electrochem. Soc.* **135**, pp. 904–906 (1988).

137. C. T. Kiers and A. Vos, The Nature of the S–S Bonds in Different Compounds II. The Crystal and Molecular Structure of Diphenyl Disulfone, *Rec. Trav. Chim. Pays-Bas* **91**, pp. 126–132 (1972).

138. N. T. Tennent, S. R. Su, C. A. Poffenberger and A. Wojcicki, Synthesis of a Transition Metal-Dithionite Complex, $(\eta^5-C_5H_5)(CO)_2Fe-S(O)_2S(O)_2-Fe(CO)_2(eta^5-C_5H_5)$, *J. Organomet. Chem.* **102**, pp. C46–C48 (1975).

139. P. Reich-Rohrwig, A. C. Clark, R. L. Downs and A. Wojcicki, Reaction of Sodium-η^5-Cyclopentadienyldicarbonylferrate (0) with Sulfur Dioxide. Synthesis of Iron–Sulfur Dioxide Complexes, *J. Organomet. Chem.* **145**, pp. 57–68 (1978).

140. C. A. Poffenberger, N. H. Tennent and A. Wojcicki, Synthesis, Characterization and Some Reactions of Organometallic Complexes Containing Bridging Dithionite Ligand, *J. Organomet. Chem.* **191**, pp. 107–121 (1980).

141. Q. Zhang, X. Lu, R.-B. Huang and L.-S. Zheng, Transition-State-Like Aromaticity in the Inorganic Ions $Se_2I_4^{2+}$ and $S_2O_4^{2-}$, *Inorg. Chem.* **45**, pp. 2457–2460 (2006).

142. M. R. Truter, An Accurate Determination of the Structure of Sodium Hydroxymethanesulfinate (Rongalite), *J. Chem. Soc.* pp. 3064–3072 (1955).

143. M. R. Truter, A Detailed Refinement of the Crystal Structure of Sodium Hydroxymethanesulfinate (Rongalite), *J. Chem. Soc.* pp. 3400–3406 (1962).

144. J. S. Edgar, On the Nature of Sodium Hydroxymethanesulfinate in Aqueous Solution, *Phosph. Sulfur* **2**, pp. 181–184 (1976).

145. S. Sato, S. Higuchi and S. Tanaka, Structural Examination of "Sodium Formaldehyde Sulfoxylate" by Infrared and Raman Spectroscopy, *Nippon Kagaku Kaishi* **1984**, pp. 1151–1157 (1984).

146. N. Masciocchi, C. Rigamonti and A. Maspero, Poly [di-μ-hydroxymethanesulfinato-zinc(II)], *Acta Cryst. E* **61**, pp. 2683–2685 (2005).

147. R. A. L. Sullivan and A. Hargreaves, The Crystal and Molecular Structure of Thiourea Dioxide, *Acta Cryst.* **15**, pp. 675–682 (1962).

148. I.-C. Chen and Y. Wang, Reinvestigation of the Structure of Thiourea *S,S*-Dioxide, $CH_4N_2O_2S$, *Acta Cryst.* **C40**, pp. 1937–1938 (1984).

149. Y. Wang, N.-L. Chang and C.-T. Pai, Charge Density Study of Thiourea *S,S*-Dioxide, $CH_4N_2O_2S$, *Inorg. Chem.* **29**, pp. 3256–3259 (1990).

150. J. S. Song, E. H. Kim, S. K. Kang, S. S. Yun, I.-H. Suh, S.-S. Choi, S. Lee and W. P. Jensen, The Structure and *Ab Initio* Studies of Thiourea Dioxide, *Bull. Korean Chem. Soc.* **17**, pp. 201–205 (1996).

151. J. F. Ojo, J. L. Petersen, A. Otoikhian and R. H. Simoyi, Organosulfur Oxoacids. Part 1. Synthesis, Structure, and Reactivity of Dimethylaminoiminomethanesulfinic Acid, *Can. J Chem.* **84**, pp. 825–830 (2006).

152. C. R. Lee, T.-H. Tang, L. Chen and Y. Wang, A Combined Experimental and Theoretical Electron Density Study of Infra- and Intermolecular Interactions in Thiourea S,S′-Dioxide, *Chem. Eur J.* **9**, pp. 3112–3121 (2003).

153. M. K. Denk, K. Hatano and A. J. Lough, Synthesis and Characterization of a Carbene SO_2 Adduct — New Insights into the Structure and Bonding of Thioure S,S′-Dioxides, *Eur J. Inorg. Chem.* **1**, pp. 224–231 (2003).

154. Z. Kis, S. V. Makarov and R. Silaghi-Dumitrescu, Computational Investigations on the Electronic Structure and Reactivity of Thiourea Dioxide: Sulfoxylate Formation, Tautomerism and Dioxygen Liberation, *J. Sulfur Chem.* **31**, pp. 27–39 (2010).

155. S. V. Makarov, C. Mundoma, J. H. Penn, J. L. Petersen, S. A. Svarovsky and R. H. Simoyi, Structure and Stability of Aminoiminomethanesulfonic Acid, *Inorg. Chim. Acta* **286**, pp. 149–154 (1999).

156. K. Peng, W. Yang and W. Zhou, Theoretical Study on Interactions between Thiourea S-Monoxide and Water, *Int. J. Quant. Chem.* **109**, pp. 811–818 (2009).

157. S. V. Makarov, Reactivity of Sulfur-Containing Reducing Agents, in *Dr. Sci. Thesis.* Ivanovo, Russia, p. 260 (2000).

158. G. Y. Fang, L. N. Xu, X. H. Li, S. Wang, J. J. Lin and W. H. Zhu, Theoretical Analysis on the Structure and Properties of Thiourea Dioxide Crystal, *Mol. Simul.* **33**, pp. 975–978 (2007).

159. K. Peng, L. Jin, W. Zhou, W. Yang and M. Li, The Investigation on the Relative Stability of Different Clusters of Thiourea Dioxide in Water Using Gas Phase Quantum Chemical Calculations, *Int. J. Quant. Chem.* **109**, pp. 1368–1375 (2009).

160. J. Shao, X. Liu, P. Chen, Q. Wu, X. Zheng and K. Pei, Investigations of TDO in Aqueous Phase by Density Functional Theory, UV Absorption, and Raman Spectroscopy, *J. Phys. Chem. A* **118**, pp. 3168–3174 (2014).

161. S. V. Makarov and E. V. Kudrik, Tautomerization of Thiourea Dioxide in Aqueous Solution, *Russ. Chem. Bull.* **50**, pp. 203–205 (2001).

162. A. J. Davidson, D. R. Allan, F. P. A. Fabbiani, D. J. Francis, W. G. Marshall, C. R. Pulham and J. E. Warren, Isostructural Transformation and Polymorphism of Thiourea Dioxide at High Pressure, *Acta Cryst. A* **61**, pp. C462–C463 (2005).

163. Q. Wang, T. Yan, K. Wang, H. Zhu, Q. Cui and B. Zou, Pressure Induced Reversible Phase Transition in Thiourea Dioxide Crystal, *J. Chem. Phys.* **142**, pp. 244701–244707 (2015).

164. Sodium-Dithionite, *CAS No. :7775-14-6.* OECD SIDS, Berlin (2004).

165. R. L. Ake, D. M. Oglesby and W. P. Kilroy, A Spectroscopic Investigation of Lithium Dithionite and the Discharge Products of a Li/SO_2 Cell, *J. Electrochem. Soc.: Electrochem. Sci. Technol.* **131**, pp. 968–974 (1984).

166. G. Vázquez, E. Alvarez, R. Varela, A. Cancela and J. M. Navaza, Density and Viscosity of Aqueous Solutions of Sodium Dithionite, Sodium

Hydroxide, Sodium Dithionite + Sucrose, and Sodium Dithionite + Sodium Hydroxide + Sucrose from 25 °C to 40 °C, *J. Chem. Eng. Data* **41**, pp. 244–248 (1996).

167. L. Erdey, J. Simon, S. Gál and G. Liptay, Thermoanalytical Properties of Analytical-Grade Reagents-IVA. Sodium Salts, *Talanta* **13**, pp. 67–80 (1966).

168. B. Flaherty and J. M. Bather, Thermal Decomposition of Sodium Dithionite, *J. Appl. Chem. Biotechnol.* **21**, pp. 236–237 (1971).

169. R. W. Merriman, The Estimation of Sodium Hydrosulfite, *J. Soc. Chem. Ind.* **42**, pp. 290–292 (1923).

170. R. Wollak, Zur jodometrischen Analyse eines Gemenges von Hydrosulfit, Sulfit und Thiosulfat, *Fresenius J. Anal. Chem.* **80**, pp. 1–4 (1930).

171. H. Zocher and H. Saechtling, Kritische Untersuchungen zur Analyse von Hyposulfit-Prparaten, *Fresenius J. Anal. Chem.* **117**, pp. 392–400 (1939).

172. L. Szekeres, Zur Analyse der Dithionite, *Fresenius J. Anal. Chem.* **203**, pp. 178–180 (1964).

173. J. P. Danehy and C. W. Zubritsky, Iodometric Method for the Determination of Dithionite, Bisulfite, and Thiosulfate in the Presence of Each Other and Its Use in Following the Decomposition of Aqueous Solutions of Sodium Dithionite, *Anal. Chem.* **46**, pp. 391–395 (1974).

174. W. P. Kilroy, Thiosulphate in the Wollak Procedure — Analysis and Kinetics in Acidic Formaldehyde Medium, *Talanta* **25**, pp. 359–362 (1978).

175. W. P. Kilroy, A Revised Method, and Errors in the Determination of Thiosulphate by the Wollak Method, *Talanta* **26**, pp. 111–115 (1979).

176. W. P. Kilroy, Determination of Dithionite, Thiosulphate, and Sulphite in the Presence of Alkali and /or Cyanide, *Talanta* **27**, pp. 343–347 (1980).

177. W. P. Kilroy, Analysis of Mixtures of Sulphide, Thiosulphate, Dithionite and Sulphide, *Talanta* **30**, pp. 419–422 (1983).

178. D. C. De Groot, Titrimetric Determination of Sodium Dithionite with Potassium Hexacyanoferrate(III) as Reagent and Methylene Blue as Indicator, *Fresenius J. Anal. Chem.* **229**, pp. 335–339 (1967).

179. P. Westbroek, G. Priniotakis and P. Kiekens, *Analytical Electrochemistry in Textiles.* Woodhead Publ., Cambridge, England (2005).

180. W. Furness, The Analysis of Commercial Hydrosulphites and Related Compounds by Polarographic Methods, *J. Soc. Dyers Colour.* **66**, pp. 270–277 (1950).

181. A. D. Broadbent and F. Peter, Polarographic Analysis of Sodium Dithionite, *J. Soc. Dyers Colour.* **82**, pp. 264–267 (1966).

182. V. Cermák and M. Smutek, Mechanism of Decomposition of Dithionite in Aqueous Solutions, *Coll. Czech. Chem. Commun.* **40**, pp. 3241–3264 (1975).

183. L. M. de Carvalho and G. Schwedt, Polarographic Determination of Dithionite and Its Decomposition Products: Kinetic Aspects, and Analytical Applications, *Anal. Chim. Acta* **436**, pp. 293–300 (2001).

184. L. M. de Carvalho and G. Schwedt, Electrochemical Behavior of Dithionite in Formaldehyde Aqueous Solution and Its Analytical Application, *Electroanalysis* **13**, pp. 596–600 (2001).

185. A. M. Alizadeh, M. Mohseni, A. A. Zamani and K. Kamali, Polarographic Determination of Sodium Hydrosulfite Residue (Dithionite) in Sugar and Loaf Sugar, *Food Analyt. Met.* **8**, pp. 483–488 (2015).
186. F. Govaert, E. Temmerman and P. Kiekens, Development of Voltammetric Sensors for the Determination of Sodium Dithionite and Indanthrene/Indigo Dyes in Alkaline Solutions, *Anal. Chim. Acta* **385**, pp. 307–314 (1999).
187. E. Gasana, P. Westbroek, E. Temmerman, H. P. Thun and P. Kiekens, A Wall-jet Disc Electrode for Simultaneous and Continuous On-line Measurement of Sodium Dithionite, Sulfite and Indigo Concentrations by Means of Multistep Chronoamperometry, *Anal. Chim. Acta* **486**, pp. 73–83 (2003).
188. A. Salimi, M. Roushani and R. Hallaj, Micromolar Determination of Sulfur Oxoanions and Sulfide at a Renewable Sol-Gel Carbon Ceramic Electrode Modified with Nickel Powder, *Electrochim. Acta* **51**, pp. 1952–1959 (2006).
189. L. M. De Carvalho and G. Schwedt, Sulfur Speciation by Capillary Zone Electrophoresis. Determination of Dithionite and Its Decomposition Products Sulfite, Sulfate and Thiosulfate in Commercial Bleaching Agents, *J. Chromatogr. A* **1099**, pp. 185–190 (2005).
190. M. Nováková, L. Krivánková, M. Bartos, V. Urbanová and K. Vytras, Isotachophoretic Determination of Hydrosulfite and Metabisulfite in Technical Samples, *Talanta* **74**, pp. 183–189 (2007).
191. J. Weiß and M. Göbl, Analyse anorganischer Schwefelverbindungen mit Hilfe der Ionen-Chromatographie, *Fresenius J. Anal. Chem.* **320**, pp. 439–440 (1985).
192. T. James, A. Apblett and N. F. Materer, Rapid Quantification of Sodium Dithionite by Ion Chromatography, *Ind. Eng. Chem. Res.* **51**, pp. 7742–7746 (2012).
193. T. James, C. Cannon, A. Apblett and N. F. Materer, Sodium Dithionite Purity and Decomposition Products in Solid Samples Spanning 50 Years, *Phosphorus Sulfur Silicon* **190**, pp. 158–169 (2015).
194. K. J. Stutts, Liquid Chromatographic Assay of Dithionite and Thiosulfate, *Anal. Chem.* **59**, pp. 543–544 (1987).
195. R. Steudel and V. Münchow, Determination of Dithionite ($S_2O_4^{2-}$) and Hydroxymethanesulphinate ($HOCH_2SO_2^-$,Rongalite) by Ion-Pair Chromatography, *J. Chromatogr. A* **623**, pp. 174–177 (1992).
196. V. Münchow and R. Steudel, The Decomposition of Aqueous Dithionite and Its Reactions with Polythionates $S_nO_6^{2-}$ ($n = 3$–5) Studied by Ion-Pair Chromatography, *Z. Anorg. Allg. Chem.* **620**, pp. 121–126 (1994).
197. B. Meyer, M. Ospina and L. B. Peter, Raman Spectrometric Determination of Oxysulfur Anions in Aqueous Systems, *Anal. Chim. Acta* **117**, pp. 301–311 (1980).
198. L. M. De Carvalho and G. Schwedt, Spectrophotometric Determination of Dithionite in Household Commercial Formulations Using Naphtol Yellow S, *Microchim. Acta* **138**, pp. 83–87 (2002).
199. W. P. Kilroy, Anaerobic Decomposition of Sodium Dithionite in Alkaline Solution, *J. Inorg. Nucl. Chem.* **42**, pp. 1071–1073 (1980).
200. F. Feigl and V. Anger, *Spot Tests in Inorganic Analysis*, 6th edn. Elsevier Sci., Amsterdam (1972).

201. W. Saus, D. Knittel and E. Schollmeyer, Voltammetric Determination of Reducing Agents and Dyestuffs in Textile Printing Pastes, *Fresenius J. Anal. Chem.* **338**, pp. 912–916 (1990).

202. S. Wang, Q. Gao and J. Wang, Thermodynamic Analysis of Decomposition of Thiourea and Thiourea Oxides, *J. Phys. Chem. B* **109**, pp. 17281–17289 (2005).

203. S. V. Makarov, V. V. Kolesnik, K. M. Dunaeva, V. V. Budanov and V. I. Spitsin, Thermal Decomposition of Rongalite and Thiourea Dioxide, *Izv. Vyssh. Uchebn. Zaved. Khim. Khim. Tekhnol.* **26**, pp. 146–150 (1983).

204. E. Mambo and R. H. Simoyi, Kinetics and Mechanism of the Complex Oxidation of Aminoiminomethanesulfinic Acid by Iodate in Acidic Media, *J. Phys. Chem.* **97**, pp. 13662–13667 (1993).

205. S. V. Makarov, C. Mundoma, J. H. Penn, S. A. Svarovsky and R. H. Simoyi, New and Surprising Experimental Results from the Oxidation of Sulfinic and Sulfonic Acids, *J. Phys. Chem. A* **102**, pp. 6786–6792 (1998).

206. Q. Gao, G. Wang, Y. Sun and I. R. Epstein, Simultaneous Tracking of Sulfur Species in the Oxidation of Thiourea by Hydrogen Peroxide, *J. Phys. Chem. A* **112**, pp. 5771–5773 (2008).

207. S. A. Svarovsky, R. H. Simoyi and S. V. Makarov, Reactive Oxygen Species in Aerobic Decomposition of Thiourea Dioxides, *J. Chem. Soc. Dalton Trans.* **29**, pp. 511–514 (2000).

208. Y. Y. Kharitonov and I. V. Prokofeva, Infrared Spectra and Structure of Thiourea Dioxide, *Dokl. Akad. Nauk SSSR* **162**, pp. 829–832 (1965).

209. J. Meyer, Zur Kenntnis der Hydroschwfligen Säure, *Z. Anorg. Chem.* **34**, pp. 43–61 (1903).

210. K. Jellinek, Über die Herstellung von Reinem Hydrosulfit und über das System Hydrosulfit-Wasser, *Z. Anorg. Chem.* **70**, pp. 93–134 (1911).

211. I. M. Kolthoff and C. S. Miller, The Reduction of Sulfurous Acid (Sulfur Dioxide) at the Dropping Mecury Electrode, *J. Am. Chem. Soc.* **63**, pp. 2818–2821 (1941).

212. M. S. Spencer, Chemistry of Sodium Dithionite. Part 1. Kinetics of Decomposition in Aqueous Bisulfite Solution, *Trans. Faraday Soc.* **63**, pp. 2510–2515 (1967).

213. L. Burlamacchi, G. Guarini and E. Tiezzi, Mechanism of Decomposition of Sodium Dithionite in Aqueous Solution, *Trans. Faraday. Soc.* **65**, pp. 496–502 (1969).

214. W. J. Lem and M. Wayman, Decomposition of Aqueous Dithionite. Part I. Kinetics of Decomposition of Aqueous Sodium Dithionite, *Can. J. Chem.* **48**, pp. 776–781 (1970).

215. R. G. Rinker, S. Lynn, D. M. Mason and W. H. Corcoran, Kinetics and Mechanism of Thermal Decomposition of Sodium Dithionite in Aqueous Solution, *Ind. Eng. Chem. Fundam.* **4**, pp. 282–288 (1965).

216. M. Wayman and W. J. Lem, Decomposition of Aqueous Dithionite. Part II. A Reaction Mechanism for the Decomposition of Aqueous Sodium Dithionite, *Can. J. Chem.* **48**, pp. 782–787 (1970).

217. M. Smutek and V. Cermák, Mathematical Treatment of the Model of Decomposition of Dithionite in Aqueous Solutions, *Coll. Czech. Chem. Commun.* **40**, pp. 3265–3280 (1975).

218. V. Cermák, Polarografická Studie Dithionicitanu Sodného, *Chem. Zvesti* **8**, pp. 714–720 (1954).

219. D. A. Holman and D. W. Bennett, A Multicomponent Kinetics Study of the Anaerobic Decomposition of Aqueous Sodium Dithionite, *J. Phys. Chem.* **98**, pp. 13300–13307 (1994).

220. K. M. Kovács and G. Rábai, Large Amplitude pH Oscillations in the Hydrogen Peroxide — Dithionite Reaction in a Flow Reactor, *J. Phys. Chem. A* **105**, pp. 9183–9187 (2001).

221. K. M. Kovács and G. Rábai, Mechanism of the Oscillatory Decomposition of the Dithionite Ion in a Flow Reactor, *Chem. Commun.* **38**, pp. 790–791 (2002).

222. C. R. Williams and D. N. Harpp, Sulfur Extrusion Reactions — Scope and Mechanistic Aspects, *Sulf. Rep.* **10**, pp. 103–191 (1990).

223. K. J. Miller, K. F. Moschner and K. T. Potts, Theoretical Study of Thieno [3,4-c]thiophene, a Nonclassical Thiophene System, *J. Am. Chem. Soc.* **105**, pp. 1705–1712 (1983).

224. D. A. Holman, A. W. Thompson, D. W. Bennett and J. D. Otvos, Quantitative Determination of Sulfur–Oxygen Anion Concentrations in Aqueous Solution: Multicomponent Analysis of Attenuated Total Reflectance Infrared Spectra, *Anal. Chem.* **66**, pp. 1378–1384 (1994).

225. T. Mizoguchi, Y. Takei and T. Okabe, The Chemical Behavior of Low Valence Sulfur Compounds. X. Disproportionation of Thiosulfate, Trithionate, Tetrathionate and Sulfite under Acidic Conditions, *Bull. Chem. Soc. Jpn.* **49**, pp. 70–75 (1976).

226. R. E. Connick, T. M. Tam and E. V. Deuster, Equilibrium Constant for the Dimerization of Bisulfite Ion to Form $S_2O_5^{2-}$, *Inorg. Chem.* **21**, pp. 103–107 (1982).

227. P. E. DePoy and D. M. Mason, Periodicity in the Chemically Reacting Systems. A Model for the Kinetics of the Decomposition of $Na_2S_2O_4$, *Faraday Symp. Chem. Soc.* **9**, pp. 47–54 (1974).

228. C. W. Rushing, R. C. Thompson and Q. Gao, General Model for the pH Dynamics in the Oxidation of Sulfur(-II) Species, *J. Phys. Chem. A* **104**, pp. 11561–11565 (2000).

229. C. N. Konidari and M. I. Karayannis, Kinetic and Mechanistic Study of the Reduction of 2,6-Dichlorophenolindophenol by Dithionites, *Anal. Chim. Acta* **224**, pp. 199–210 (1989).

230. M. W. Lister and R. C. Garvie, Sodium Dithionite, Decomposition in Aqueous Solution and in the Solid State, *Can. J. Chem.* **37**, pp. 1567–1574 (1959).

231. S. V. Makarov, Y. V. Polenov and V. V. Budanov, Polarographic Study of Rongalite Solution, *Zh. Neorg. Khim.* **29**, pp. 2456–2460 (1984).

232. Q. Gao, B. Liu, L. Li and Y. Wang, Oxidation and Decomposition Kinetics of Thiourea Oxides, *J. Phys. Chem. A* **111**, pp. 872–877 (2007).

233. H. Bassett and R. G. Durrant, CXCVII — The Inter-Relationship on the Sulphur Acids, *J. Chem. Soc.* pp. 1401–1468 (1927).
234. M. M. Nicloux, Sur L'Oxydation de l'Hydrosulfite de Sodium par L'oxygene, *Compt. Rend.* **196**, pp. 616–634 (1933).
235. S. Lynn, *Radical Diffusion in a Turbulent Air System. Absorption of Light by the System Nitric Acid–Nitrogen Dioxide–Water. Ionization in Solutions of Nitrogen Dioxide in Nitric Acid from Optical Absorbance Measurements. Kinetics of the Decomposition of Sodium Dithionite. The Determination of Chromium by Oxidation in the Presence of Silver Nitrate.*, Dissertation (Ph.D). California Institute of Technology, Pasadena, Ca (1954).
236. R. G. Rinker, T. P. Gordon, D. M. Mason, R. R. Sakaida and W. H. Corcoran, Kinetics and Mechanism of the Air Oxidation of the Dithionite Ion ($S_2O_4^=$) in Aqueous Solution, *J. Phys. Chem.* **64**, pp. 573–581 (1960).
237. D. Lambeth and G. Palmer, The Kinetics and Mechanism of Reduction of Electron Transfer Proteins and Other Compounds of Biological Interest by Dithionite, *J. Biol. Chem.* **248**, pp. 6095–6103 (1973).
238. C. Creutz and N. Sutin, Kinetics of the Reactions of Sodium Dithionite with Dioxygen and Hydrogen Peroxide, *Inorg. Chem.* **13**, pp. 2041–2043 (1974).
239. J. A. Morello, M. R. Craw, H. P. Constantine and R. E. Forster, Rate of Reaction of Dithionite Ion with Oxygen in Aqueous Solution, *J. Appl. Physiol.* **19**, pp. 522–525 (1964).
240. Z. Tao, J. Goodisman and A.-K. Souid, Oxygen Measurement via Phosphorescence: Reaction of Sodium Dithionite with Dissolved Oxygen, *J. Phys. Chem. A* **112**, pp. 1511–1518 (2008).
241. D. K. Singh, R. N. Sharma and R. D. Srivastava, Kinetics of Oxidation of Sodium Dithionite by Flow Thermal Method, *AIChE J.* **24**, pp. 232–237 (1978).
242. M. Kawagoe and C. W. Robinson, The Kinetics of Dissolved Oxygen Reaction in Aqueous Sodium Dithionite Solutions, *Can. J. Chem. Eng.* **59**, pp. 471–474 (1981).
243. S. Fukushima, A. Uyama, Y. Yamaguchi, E. Tsuji and S. Mezaki, Oxygen Absorption into Sodium Dithionite Solution, *J. Chem. Eng. Jap.* **11**, pp. 283–289 (1978).
244. H. Hikita, H. Ishikawa, N. Sakamoto and N. Esaka, Kinetics of Absorption of Oxygen in Aqueous Alkaline Solutions of Sodium Dithionite, *Chem. Eng. Sci.* **33**, pp. 392–396 (1978).
245. A. S. Jhaveri and M. M. Sharma, Absorption of Oxygen in Aqueous Alkaline Solutions of Sodium Dithionite, *Chem. Eng. Sci.* **23**, pp. 1–8 (1968).
246. L. Forlani, C. Ioppolo, E. Antonini, C. J. Martin and M. A. Marini, Calorimetric Studies of Oxyhemoglobin Dissociation. I. Reaction of Sodium Dithionite with Oxygen, *J. Inorg. Biochem.* **20**, pp. 147–155 (1984).
247. F. Camacho, M. P. Páez, G. Blázquez and J. M. Garrido, Oxygen Absorption in Alkaline Sodium Dithionite Solutions, *Chem. Eng. Sci.* **47**, pp. 4309–4314 (1992).
248. F. Camacho, M. P. Páez, G. Blázquez, M. C. Jiménez and M. Fernández, Influence of pH on the Oxygen Absorption Kinetics in Alkaline Sodium Dithionite Solutions, *Chem. Eng. Sci.* **50**, pp. 1181–1186 (1995).

249. F. Camacho, M. P. Páez, M. C. Jiménez and M. Fernández, Application of the Sodium Dithionite Oxidation to Measure Oxygen Transfer Parameters, *Chem. Eng. Sci.* **52**, pp. 1387–1391 (1997).

250. A. A. Shaikh and S. M. J. Zaidi, Kinetics of Oxygen Absorption in Aqueous Sodium Dithionite Solutions, *J. Chem. Technol. Biotechnol.* **56**, pp. 139–145 (1993).

251. E. Skavas and T. Hemmingsen, A Study of the Interference from Sulphide and Dithionite on Electrochemical Sulphite Analyses on Platinum, *Int. J. Electrochem. Sci.* **2**, pp. 203–215 (2007).

252. R. E. Huie, C. L. Clifton and N. Altstein, A Pulse Radiolysis and Flash Photolysis Study of the Radicals SO_2^-, SO_3^-, SO_4^-, SO_5^-, *Radiat. Phys. Chem.* **33**, pp. 361–370 (1989).

253. S. V. Makarov, C. Mundoma, S. A. Svarovsky, X. Shi, P. M. Gannett and R. H. Simoyi, Reactive Oxygen Species in the Aerobic Decomposition of Sodium Hydroxymethanesulfinate, *Arch. Biochem. Biophys.* **367**, pp. 289–296 (1999).

254. D. Liu, H. Lin, C. Tang and J. Du, Sulfur Dioxide: A Novel Gaseous Signal in the Regulation of Cardiovascular Functions, *Mini-Rev. Med. Chem.* **10**, pp. 1039–1045 (2010).

255. X.-B. Wang, J.-B. Du and H. Cui, Sulfur Dioxide, a Double-faced Molecule in Mammals, *Life Sci.* **98**, pp. 63–67 (2014).

256. C. Mottley, L. S. Harman and R. P. Mason, Microsomal Reduction of Bisulfite (Aqueous Sulfur Dioxide) — Sulfur Dioxide Anion Free Radical Formation by Cytochrome P-450, *Biochem. Pharmacol.* **34**, pp. 3005–3008 (1985).

257. T. V. Mishanina, M. Libiad and R. Banerjee, Biogenesis of Reactive Sulfur Species for Signaling by Hydrogen Sulfide Oxidation Pathways, *Nat. Chem. Biol.* **11** pp. 457–464 (2015).

258. T. N. Das, R. E. Huie, P. Neta and S. Padmaja, Reduction Potential of the Sulfhydryl Radical: Pulse Radiolysis and Laser Flash Photolysis Studies of the Formation and Reactions of SH and $HSSH^-$ in Aqueous Solutions, *J. Phys. Chem. A* **103** pp. 5221–5226 (1999).

259. W. G. Hodgson, A. Neaves and C. A. Parker, Detection of Free Radicals in Sodium Dithionite by Paramagnetic Resonance, *Nature* **178**, p. 489 (1956).

260. R. G. Rinker, T. P. Gordon, D. M. Mason and W. H. Corcoran, The Presence of the SO_2^- Radical Ion in Aqueous Solutions of Sodium Dithionite, *J. Phys. Chem.* **63**, p. 302 (1959).

261. H. C. Clark, A. Horsfield and M. R. Symons, Unstable Intermediates. Part XII. The Radical-Ions SO_2^- and NO_2^{2-}, *J. Chem. Soc.*, pp. 7–11 (1961).

262. P. W. Atkins, A. Horsfield and M. R. Symons, Oxides and Oxyions of the Non-metals. Part VIII. SO_2^- and ClO_2, *J. Chem. Soc.*, pp. 5220–5225 (1964).

263. G. Mayhew, The Redox Potential of Dithionite and SO_2^- from Equilibrium Reactions with Flavodoxins, Methyl Viologen and Hydrogen Plus Hydrogenase, *Eur. J. Biochem.* **85**, pp. 535–547 (1978).

264. P. Neta, R. E. Huie and A. Harriman, One-Electron-Transfer Reactions of the Couple $SO_2/\cdot SO_2^-$ in Aqueous Solutions. Pulse Radiolytic and Cyclic Voltammetric Studies, *J. Phys. Chem.* **91**, pp. 1606–1611 (1987).

265. T. E. Eriksen, Pulse Radiolytic Investigation of the SO_2^- Radical Ion, *Radiochem. Radioanal. Lett.* **22**, pp. 33–40 (1975).

266. P. Neta and R. E. Huie, Free-Radical Chemistry of Sulfite, *Environ. Health Persp.* **64**, pp. 209–217 (1985).

267. T. Ozawa and T. Kwan, ESR Studies on the Reactive Character of the Radical Anions, SO_2^-, SO_3^- and SO_4^- in Aqueous Solution, *Polyhedron* **2**, pp. 1019–1023 (1983).

268. P. C. Harrington and R. G. Wilkins, Reduction of Methemerythrin by e_{aq}^- and CO_2^- from Pulse Radiolysis Studies, *J. Biol. Chem.* **254**, pp. 7505–7508 (1979).

269. W. A. Waters, Alkylpentacyanocobalt Nitroxide and Primary Alkyl Nitroxide Radical-anions, *J. Chem. Soc. Chem. Commun.* **8**, pp. 1087–1088 (1972).

270. D. Mulvey and W. A. Waters, Some Nitrosylsulphinate Radical-anions •$RN(SO_2^-)O$, *J. Chem. Soc. Perkin. Trans. II* pp. 772–773 (1974).

271. J. C. Evans, S. K. Jackson and C. C. Rowlands, An Electron Spin Resonance Study of Radicals from Chloramine-T-2: Spin Trapping of Photolysis Products of Chloramine-T at Alkaline pH, *Tetrahedron* **41**, pp. 5195–5200 (1985).

272. D. Rehorek, C. M. DuBose and E. G. Janzen, On the Spin Trapping of Sulfur Dioxide Anion-radicals (•SO_2^-) by Nitrones, *J. Prakt. Chem.* **333**, pp. 321–325 (1991).

273. K. Stolze, D. R. Duling and R. P. Mason, Spin Adducts Formed from Nitroso Spin Traps and Dithionite, *J. Chem. Soc. Chem. Commun.* **24**, pp. 268–270 (1988).

274. C. Mottley, R. P. Mason, C. F. Chignell, K. Sivarajah and T. E. Eling, The Formation of Sulfur Trioxide Radical Anion during the Prostaglandin Hydroperoxidase-catalyzed Oxidation of Bisulfite (Hydrated Sulfur Dioxide), *J. Biol. Chem.* **257**, pp. 5050–5055 (1982).

275. M. Yonemura, T. Sekine and H. Ueda, SO_2^- Doping of $SrTiO_3$ and $SrZrO_3$ by Cyclic (CS_2–O_2) Processing, *J. Phys. Chem.* **90**, pp. 3003–3005 (1986).

276. M. A. Yu, T. Egawa, S.-R. Yeh, D. L. Rousseau and G. J. Gerfen, EPR Characterization of Ascorbyl and Sulfur Dioxide Anion Radicals Trapped During the Reaction of Bovine Cytochrome c Oxidase with Molecular Oxygen, *J. Magn. Res.* **203**, pp. 213–219 (2010).

277. E. G. Janzen, Electron Spin Resonance Study of the •SO_2^- Formation in the Thermal Decomposition of Sodium Dithionite, Sodium and Potassium Metabisulfite, and Sodium Hydrogen Sulfite, *J. Phys. Chem.* **76**, pp. 157–162 (1972).

278. E. Potteau, E. Levillain and J.-P. Lelieur, Spectroscopic Characterization of the Blue and Red Complexes in Reduced SO_2 Solutions in HMPA, *New J. Chem.* **23**, pp. 1117–1124 (1999).

279. O. Ito, S. Kuwashima and M. Matsuda, Absorption Spectrum of the Aggregated Radical Anion of Sulfur Dioxide Produced by γ-Irradiation, *Bull. Soc. Jap.* **49**, pp. 327–328 (1976).

280. D. Knittel, Electrolytically Generated Sulfur Dioxide Anion Radical •$S_2O_4^-$, Its Absorption Coefficient and Some of Its Decay Reactions, *J. Electroanal. Chem.* **195**, pp. 345–356 (1985).

281. W. D. Harrison, J. B. Gill and D. C. Goodall, Reactions in Mixed Non-aqueous Systems Containing Sulphur Dioxide. Part 2. The Dissolution of Transition Metals in the Binary Mixture Dimethyl Sulphoxide — Sulphur Dioxide, and Ion-pair Formation Involving the Sulphoxylate Radical Ion in Mixed Solvents Containing Sulphur Dioxide, *J. Chem. Soc. Dalton Trans.* **8**, pp. 847–850 (1979).

282. E. M. Burgess, U. Zoller and R. L. J. Burger, Conversion of Thiourea Dioxide to Dioxygen, *J. Am. Chem. Soc.* **106**, pp. 1128–1130 (1984).

283. C.-P. Zhang, Q.-Y. Chen, Y. Guo, J.-C. Xiao and Y.-C. Gu, Progress in Fluoroalkylation of Organic Compounds via Sulfinatodehalogenation Initiation System, *Chem. Soc. Rev.* **41**, pp. 4536–4559 (2012).

284. W.-Y. Huang and F.-H. Wu, The Sulfinatodehalogenation Reaction, *Isr. J. Chem.* **39**, pp. 167–170 (1999).

285. C. Ni, M. Hu and J. Hu, Good Partnership between Sulfur and Fluorine: Sulfur-Based Fluorination and Fluoroalkylation Reagents for Organic Synthesis, *Chem. Rev.* **115**, pp. 765–825 (2015).

286. B.-N. Huang, W.-Y. Huang and C.-M. Hu, Deiodosulfination of Perfluoroalkyl Iodides and a New Method for the Synthesis of Perfluorosulfonic Acids, *Acta Chim. Sin.* **39**, p. 483 (1981).

287. W.-Y. Huang, B.-N. Huang and C.-M. Hu, Studies on Deiodo-Sulfination. Part 1. Studies on the Deiodo-Sulfination of Perfluoroalkyl Iodides, *J. Fluorine Chem.* **23**, pp. 193–204 (1983).

288. W.-Y. Huang, B.-N. Huang and W. Wang, Studies on Sulfinatodehalogenation III. The Sulfinatodeiodination of Primary Perfluoroalkyl Iodides and α, ω-Perfluoroalkylene Diiodides by Sodium Dithionite, *Acta Chim. Sin.* **3**, pp. 252–256 (1985).

289. W.-Y. Huang, A. Haas and M. Lieb, A New Method for the Preparation of Perfluorocarboxylic Acids, *J. Fluorine Chem.* **36**, pp. 49–62 (1987).

290. W.-Y. Huang and J.-T. Liu, Studies on Sulfinatodehalogenation. XVII. The Sulfinatodehalogenation of Primary Perfluoroalkyl Iodides and Bromides by Rongalite, *Chin. J. Chem.* **8**, pp. 355–357 (1990).

291. W.-Y. Huang and J.-T. Liu, Rongalite-Initiated Addition of Perfluoroalkyl Iodides to Olefins, *Chin. J. Chem.* **8**, pp. 358–360 (1990).

292. M. Tordeux, B. Langlois and C. Wakselman, Reactions of Bromotrifluoromethane and Related Halides. 8. Condensations with Dithionite and Hydroxymethanesulfinate Salts, *J. Org. Chem.* **54**, pp. 2452–2453 (1989).

293. B. J. Grady and D. C. Dittmer, Reaction of Perfluoroaryl Halides with Reduced Species of Sulfur Dioxide (HSO_2^-, SO_2^{2-}, $S_2O_4^{2-}$), *J. Fluorine Chem.* **50**, pp. 151–172 (1990).

294. B.-N. Huang and J.-T. Liu, Studies on Sulfinatodehalogenation. XXVI. Perfluoroalkylation of Nitrogen-Containing Heteroaromatic Compounds with Perfluoroalkyl Halides and Rongalite, *J. Fluorine Chem.* **64**, pp. 37–46 (1993).

295. B.-N. Huang, J.-T. Liu and W.-Y. Huang, Direct Perfluoroalkylation of Coumarins by Perfluoroalkyl Iodides in the Presence of Sodium

Hydroxymethanesulphinate (Rongalite), *J. Chem. Soc. Chem. Commun.* pp. 1781–1782 (1990).

296. B.-N. Huang, J.-T. Liu and H. W.-Y., Studies on Sulfinatodehalogenation. Part 30. Synthesis of 3-Perfluoroalkylated Coumarins, Thiocoumarins and 2-Quinolones by Direct Perfluoroalkylation with Perfluoroalkyl Iodides and Sodium Hydroxymethanesulfinate, *J. Chem. Soc. Perkin Trans.* 1, pp. 101–104 (1994).

297. C.-P. Zhang, Z.-L. Wang, C. Q.-Y., C.-T. Zhang, Y.-C. Gu and J.-C. Xiao, Generation of the CF_3 Radical from Trifluoromethylsulfonium Triflate and Its Trifluoromethylation of Styrenes, *Chem. Commun.* **47**, pp. 6632–6634 (2011).

298. G. G. Furin, Some New Aspects in the Application of Perfluoroalkyl Halides in the Synthesis of Fluorine-Containing Organic Compounds, *Russ. Chem. Rev.* **69**, pp. 491–522 (2000).

299. B. Barkakaty, Y. Takaguchi and S. Tsuboi, New Synthetic Routes towards Various α-Fluorinated Aryl Ketones and Their Enantioselective Reductions Using Baker's Yeast, *Tetrahedron* **63**, pp. 970–976 (2007).

300. W.-Y. Huang and J.-L. Zhuang, Studies on Sulfinatodehalogenation. XXIII. A New Sulfinatodehalogenation Reagent: Thiourea Dioxide, *Chin. J. Chem.* **9**, pp. 270–274 (1991).

301. C. Wakselman and M. Tordeux, Perfluoroalkylation of Anilines in the Presence of Zinc and Sulphur Dioxide, *J. Chem. Soc. Chem. Commun.* **23**, pp. 1701–1703 (1987).

302. M. Tordeux, B. Langlois and C. Wakselman, Reactions of Trifluoromethyl Bromide and Related Halides: Part 10. Perfluoroalkylation of Aromatic Compounds Induced by Sulphur Dioxide Radical Anion Precursor, *J. Chem. Soc. Perkin Trans.* 1, pp. 2293–2299 (1990).

303. C. Liu, D.-M. Shen, Z. Zeng, C.-C. Guo and Q.-Y. Chen, C-F Bond Activation by Modified Sulfinatodehalogenation: Facile Synthesis and Properties of Novel Tetrafluorobenzoporphyrins by Direct Intramolecular Cyclization and Reductive Defluorinative Aromatization of Readily Available β-Perfluoroalkylated Porphyrins, *J. Org. Chem.* **71**, pp. 9772–9783 (2006).

304. X.-J. Tang and Q.-Y. Chen, Insight into "Entrainment" in $S_{RN}1$ Reactions of 2,2-Dichloro-1,1,1-Trifluoroethane (HCFC-123) with Thiolates Initiated by $Na_2S_2O_4$, *J. Fluorine Chem.* **169**, pp. 1–5 (2015).

305. C. De Schutter, E. Pfund and T. Lequeux, Radical Conjugate Addition of Ambiphilic Fluorinated Free Radicals, *Tetrahedron* **69**, pp. 5920–5926 (2013).

306. Z. Ma and S. Ma, Stereoselective Synthesis of 2-Iodo-1-Perfluoroalkyl-2(Z)-Alkenes and E or Z-4-Perfluoroalkylmethyl-4-en-2-ynol via $Na_2S_2O_4$-Promoted Radical Addition Reaction of Perfluoroalkyl Iodides with Allenes and the Palladium-Catalyzed Kinetic Resolution with Sonogashira Coupling Reaction, *Tetrahedron* **64**, pp. 6500–6509 (2008).

307. B. R. Langlois, E. Laurent and N. Roidot, Trifluoromethylation of Aromatic Compounds with Sodium Trifluoromethanesulfinate under Oxidative Conditions, *Tetrahedron Lett.* **32**, pp. 7525–7528 (1991).

308. B. R. Langlois, E. Laurent and N. Roidot, "Pseudo-Cationic" Trifluoromethylation of Enol Esters with Sodium Trifluoromethanesulfinate, *Tetrahedron Lett.* **32**, pp. 1291–1294 (1992).

309. B. R. Langlois, E. Montegre and N. Roidot, Synthesis of S-trifluoromethyl-containing α-Amino Acids from Sodium Trifluoromethanesulfinate and Dithio-amino Acids, *J. Fluorine Chem.* **68**, pp. 63–66 (1994).

310. C. Zhang, Application of Langlois Reagent in Trifluoromethylation Reactions, *Adv. Synth. Catal.* **356**, pp. 2895–2906 (2014).

311. W.-Z. Ge, Y.-M. Wu and W.-Y. Huang, Studies on the Reactions of Silyl Enol Ether with Perfluoroorganic Compounds, *Chin. J. Chem.* **9**, pp. 527–535 (1991).

312. M. Hein, R. Miethchen and D. Swabisch, Compounds and Fluorinating Agents. Part 25. Dithionite Mediated C- and S-Perfluoroalkylations of Monosaccharide Derivatives, *J. Fluorine Chem.* **98**, pp. 55–60 (1999).

313. S. Mao, X. Fang, L. Ba and F. Wu, The Reaction of Fluorine-containing Compounds with Conjugated Dienoic Acids Initiated by Sodium Dithionite, *J. Fluorine Chem.* **128**, pp. 5–11 (2007).

314. X. Yang, Z. Wang, Y. Zhu, X. Fang, X. Yang, F. Wu and Y. Shen, Fluoroalkylation of Pent-4-en-1-ols Initiated by Sodium Dithionite to Synthesize Fluorine-Containing Tetrahydrofuran Derivatives, *J. Fluorine Chem.* **128**, pp. 1046–1051 (2007).

315. Z. Xiao, H. Hu, J. Ma, Q. Chen and Y. Guo, Radical Addition of Perfluoroalkyl Iodides to Alkenes and Alkynes Initiated by Sodium Dithionite in an Aqueous Solution in the Presence of a Novel Fluorosurfactant, *Chin. J. Chem.* **31**, pp. 939–944 (2013).

316. V. Petrik and D. Cahard, Trifluoromethylation of Ammonium Enolates, *Tetrahedron Lett.* **48**, pp. 3327–3330 (2007).

317. J. C. Rodriguez and M. Rivera, Reductive Dehalogenation of Carbon Tetrachloride Carried out by Sodium Dithionite, *Chem. Lett.* **26**, pp. 1133–1134 (1997).

318. Y. Ji, T. Bruecki, R. D. Baxter, Y. Fujiwara, I. B. Seiple, S. Su, D. G. Blackmond and P. S. Baran, Innate C–H Trifluoromethylation of Heterocycles, *Proc. Natl. Acad. Sci. USA* **108**, pp. 14411–14415 (2011).

319. T. Ishikawa, *Guanidines in Organic Synthesis, in Superbases in Organic Synthesis*, 1st edn. John Wiley & Sons, Chichester (2009).

320. P. Selig, Guanidine Organocatalysis, *Synthesis* **45**, pp. 0703–0718 (2013).

321. A. R. Katriczky and B. V. Rogovoy, Recent Developments in Guanylating Agents, *Arkivoc* **4**, pp. 49–87 (2005).

322. R. G. S. Berlinck, Natural Guanidine Derivatives, *Nat. Prod. Rep.* **16**, pp. 339–365 (1999).

323. R. G. S. Berlinck, A. E. Trindade and M. F. C. Santos, The Chemistry and Biology of Organic Guanidine Derivatives, *Nat. Prod. Rep.* **29**, pp. 1382–1406 (2012).

324. J. V. Alegre-Requena, E. Marques-Lopez and R. P. Herrera, Guanidine Motif in Biologically Active Peptides, *Aus. J. Chem.* **67**, pp. 965–971 (2014).

325. T. R. M. Pauws and B. U. W. Maes, Transition Metal-Catalyzed N-Arylations of Amidines and Guanidines, *Chem. Soc. Rev.* **41**, pp. 2463–2497 (2012).

326. T. Suhs and B. König, Synthesis of Guanidines in Solutions, *Mini-Rev. Org. Chem.* **3**, pp. 315–331 (2006).

327. C. Alonso-Moreno, A. Antinolo, F. Carillio-Hermosilla and A. Otero, Guanidines from Classical Approaches to Efficient Catalytic Syntheses, *Chem. Soc. Rev.* **43**, pp. 3406–3425 (2014).

328. W.-X. Zhang, L. Xu and Z. Xi, Recent Developments on the Synthetic Methods of Guanidines via Transition Metal Catalysis, *Chem. Commun.* **51**, pp. 254–265 (2015).

329. W. Walter, Guanidierende Wirkung der Formamidinsulfinsäure, *Angew. Chem.* **67**, pp. 275–276 (1955).

330. C. Knopp, Zur Verwendung von Aminoiminomethanesulfinsäure als Antioxidans, *Sci. Pharm.* **51**, pp. 283–290 (1983).

331. B. S. Jursic, D. Neumann and M. A., Preparation of N-Formamidinylamino Acids from Amino and Formamidinesulfinic Acids, *Sci. Pharm.* **51**, pp. 283–290 (1983).

332. S. V. Makarov, E. V. Kudrik, I. N. Terskaya and K. A. Davydov, Reaction of Thiourea Dioxide with Amines, *Russ. J. Gen. Chem.* **74**, pp. 1383–1385 (2004).

333. M. Prashad, L. Chen, O. Repic and T. J. Blackock, A New Reaction of Aminoiminomethanesulfonic Acid with Methyl Anthranilates, *Synth. Commun.* **28**, pp. 2125–2129 (1998).

334. A. E. Miller, D. J. Feeney, Y. Ma, L. Zarcone, M. A. Aziz and E. Magnuson, The Synthesis of Aminoiminoethanenitriles, 5-Aminotetrazoles, N-Cyanoguanidines, and N-Hydroxyguanidines from Aminoiminomethanesulfonic Acids, *Synth. Commun.* **20**, pp. 217–226 (1990).

335. J. Wityak, R. A. Earl, M. M. Abelman, Y. B. Bethel, B. N. Fischer, G. S. Kaufmann, C. A. Kettner, P. Ma, J. L. McMillan, L. J. Mersinger, J. Pesti, M. E. Pierce, F. W. Rankin, R. J. Chorvat and P. N. Confalone, Synthesis of Thrombin Inhibitor DuP 714, *J. Org. Chem.* **60**, pp. 3717–3722 (1995).

336. P. Mantri, D. E. Duffy and C. A. Kettner, New Asymmetric Synthesis of α-Aminoboronic Acids Containing Functionalized Side Chains, *J. Org. Chem.* **61**, pp. 5690–5692 (1996).

337. W. Han, J. C. Pelletier, L. J. Mersinger, C. A. Kettner and C. N. Hodge, 7-Azabicycloheptane Carboxylic Acid: A Proline Replacement in a Boroarginine Thrombin Inhibitor, *Org. Lett.* **1**, pp. 1875–1877 (1999).

338. T. Lu, B. E. Tomczyk, C. R. Illig, R. Bone, L. Murphy, J. Spurlino, F. R. Salemme and R. M. Soll, *In Vitro* Evaluation and Crystallographic Analysis

of a New Class of Selective, Non-Amide-Based Thrombin Inhibitors, *Bioorg. Med. Chem. Lett.* **8**, pp. 1595–1600 (1998).

339. T. Lu, R. M. Soll, C. R. Illig, R. Bone, L. Murphy, J. Spurlino, F. R. Salemme and B. E. Tomczyk, Structure-Activity and Crystallographic Analysis of a New Class of Non-Amide-Based Thrombin Inhibitor, *Bioorg. Med. Chem. Lett.* **10**, pp. 79–82 (2000).

340. E. J. Iwanowicz, J. Lin, D. G. M. Roberts, I. M. Michel and S. M. Seiter, α-Hydroxy- and α-Ketoester Functionalized Thrombin Inhibitors, *Bioorg. Med. Chem. Lett.* **2**, pp. 1607–1612 (1992).

341. J. Colanduoni and J. J. Villafranca, Labeling of Specific Lysine Residues at the Active Site of Glutamine Synthetase, *J. Biol. Chem.* **260**, pp. 15042–15050 (1985).

342. C. L. DiIanni, J. Colanduoni and J. J. Villafranca, Inactivation of *Escherichia coli* Glutamine Synthetase by Thiourea Trioxide, *Bioorg. Chem.* **14**, pp. 242–248 (1986).

343. J. G. Robertson, L. J. Sparvero and J. J. Villafranca, Inactivation and Covalent Modification of CTP Synthetase by Thiourea Dioxide, *Protein Sci.* **1**, pp. 1298–1307 (1992).

344. J. Vinsova and E. Vavrikova, Recent Advances in Drugs and Prodrugs Design of Chitosan, *Curr. Pharm. Des.* **14**, pp. 1311–1326 (2008).

345. Y. Hu, Y. Du, J. Yang, J. F. Kennedy, X. Wang and L. Wang, Synthesis, Characterization and Antibacterial Activity of Guanidinylated Chitosan, *Carbohydr. Polym.* **67**, pp. 66–72 (2007).

346. X. Zhai, P. Sun, Y. Luo, C. Ma, J. Xu and W. Liu, A Simple Way to Fabricate Cell Penetrating Peptide Analogue-Modified Chitosan Vector for Enhanced Gene Delivery, *J. Appl. Polym. Sci.* **121**, pp. 3569–3578 (2011).

347. Y. Luo, X. Zhai, C. Ma, P. Sun, Z. Fu, W. Liu and J. Xu, An Inhalable β2-Adrenoceptor Ligand-Directed Guanidilated Chitosan Carrier for Targeted Delivery of siRNA to Lung, *J. Control. Release* **162**, pp. 28–36 (2012).

348. B. O. Bachmann and C. A. Townsend, Kinetic Mechanism of the β-Lactam Synthetase of Streptomyces clavuligerus, *Biochemistry* **39**, pp. 11187–11193 (2000).

349. D. Nozawa, H. Takikawa and K. Mori, Synthesis and Absolute Configuration on Stellettadine A: A Marine Alkaloid that Induces Larval Metamorphosis in Ascidians, *Bioorg. Med. Chem. Lett.* **11**, pp. 1481–14833 (2001).

350. R. O. Dempcy, K. A. Browne and T. C. Bruice, Synthesis of the Polycation Thymidyl DNG, Its Fidelity in Binding Polyanionic DNA/RNA, and the Stability and Nature of the Hybrid Complexes, *J. Am. Chem. Soc.* **117**, pp. 6140–6141 (1995).

351. D. R. Artis, C. Brotherton-Pleiss, J. H. B. Pease, C. J. Lin, S. W. Ferla, S. R. Newman, S. Bhakta, H. Ostrelich and K. Jarnagin, Structure-Based Design of Six Novel Classes of Nonpeptide Antagonists of the Bradykinin B2 Receptor, *Bioorg. Med. Chem. Lett.* **10**, pp. 2421–2425 (2000).

352. S. Routier, J.-L. Bernier, J.-P. Catteau, J.-F. Riou and C. Bailly, Synthesis, DNA Binding, Topoisomerase II and Cytotoxicity of Two

Guanidine-Containing Anthracene-9,10-diones, *Anti-Cancer Drug Des.* **13**, pp. 407–415 (1998).

353. P. Chand, Y. S. Babu, S. Bantia, N. Chu, L. B. Cole, P. L. Kotian, W. G. Laver, J. A. Montgomery, V. P. Pathak, S. L. Petty, D. P. Shrout, D. A. Walsh and G. M. Walsh, Design and Synthesis of Benzoic Acid Derivatives as Influenza Neuramidinase Inhibitors Using Structure-Based Drug Design, *J. Med. Chem.* **40**, pp. 4030–4052 (1997).

354. A. P. Kozikowski, D. S. Dodd, J. Zaidi, Y.-P. Pang, B. Cusack and E. Richelson, Synthesis of Partial Nonpeptidic Peptide Mimetics as Potential Neurotensin Agonists and Antagonists, *J. Chem. Soc. Perkin Trans.* **1**, pp. 1615–1621 (1995).

355. M. von Itzstein, W.-Y. Wu and B. Jin, The Synthesis of 2,3-Didehydro-2,4-dideoxy-4-guanidinyl-N-acetylneuraminic Acid: A Potent Influenza Virus Sialidase Inhibitor, *Carbonhydr. Res.* **259**, pp. 301–305 (1994).

356. D. J. Aitken, D. Guillamine and H.-P. Husson, First Asymmetric Synthesis of Carnosadine, *Tetrahedron* **49**, pp. 6375–6380 (1993).

357. L. Crombie and S. R. M. Jarrett, Synthesis of Amido-Ureas and the Nature of Caracasanamide, the Hypotensive Principle of Verbesina caracasana, *J. Chem. Soc. Perkin Trans.* **1**, pp. 3179–3183 (1992).

358. W. R. Jackson, F. C. Copp, J. D. Cullen, F. J. Guyett, I. D. Rae, A. J. Robinson, H. Pothoulackis, A. K. Serelis and M. Wong, Chemical Design of Peripherally Acting Compounds, *Clin. Exp. Pharmacol. Physiol.* **19**, pp. 17–23 (1992).

359. U. Schmidt, K. Mundinger, R. Mangold and A. Lieberknecht, Lavendomycin: Total Synthesis and Assignment of Configuration, *J. Chem. Soc., Chem. Commun.* **26**, pp. 1216–1219 (1990).

360. H. Dürr, M. Goodman and G. Jung, Retro-Inverso Amide Bonds between Trifunctional Amino Acids, *J. Chem. Soc. Chem. Commun.* **31**, pp. 785–787 (1992).

361. N. S. Simpkins, *Sulphones in Organic Synthesis*, 1st edn. Pergamon Press, Oxford (1993).

362. E. Fromm, Über die Niedersten Oxyde des Schwefelwasserstoffs, *Ber. Dtsch. Chem. Ges.* **41**, pp. 3397–3425 (1908).

363. E. Wellisch, E. Gipstein and O. J. Sweeting, Use of Dithionite in the Synthesis of Sulfones and Polysulfones, *Polym. Lett.* **2**, pp. 35–37 (1964).

364. A. S. Amiri and J. M. Mellor, Photochemistry of di-α-methylenenaphtylsulphone, *J. Photochem.* **9**, pp. 571–575 (1978).

365. R. Kerber and J. Starnick, Darstellung von Symmetrischen Sulfonen durch Anlagerung von Sulfoxylsaurederivativen an Aktivierte Doppelbindungen, *Chem. Ber.* **104**, pp. 2035–2043 (1971).

366. R. Kerber and W. Gestrich, Dihydroxyarylsulfone (und chinoide folgeprodukte) durch Reaktion von Hydroxymethansulfinat (rongalit c$^{®}$) mit Chinonen, *Chem. Ber.* **106**, pp. 798–802 (1973).

367. P. Messinger and H. Greve, Cyclische Sulfon aus Mannich-Basen und Natrium-Hydroxymethansulfinat, *Arch. Pharm.* **310**, pp. 674–676 (1977).

368. A. R. Harris, A Simple, One Step Procedure for the Preparation of Dibenzyl Sulphones from Benzyl Halides, *Synth. Commun.* **18**, pp. 659–663 (1988).

369. A. Loupy, J. Sansoulet and A. R. Harris, Organic Chemistry without Solvent Improvements to the One Step Preparation of Dibenzyl Sulphones, *Synth. Commun.* **19**, pp. 2939–2946 (1989).

370. J. M. Khurana, G. Bansal and P. K. Sahoo, Reductive Coupling of Benzylic and Allylic Halides with Sodium Dithionite in DMF or HMPA, *J. Chem. Res.*, pp. 139–140 (2004).

371. M. D. Hoey and D. C. Dittmer, A Convenient Synthesis of 1,4-Dihydro-2,3-Benzoxathiin 3-Oxide, a Useful Precursor of o-Quinodimethane, *J. Org. Chem.* **56**, pp. 1947–1948 (1991).

372. S. Kotha and A. K. Ghosh, A New and Simple Synthetic Approach to Functionalized Sulphone Derivatives by the Suzuki–Miyaura Cross-Coupling Reaction, *Ind. J. Chem.* **45B**, pp. 227–231 (2006).

373. S. Kotha, P. Khedkar and A. K. Ghosh, Synthesis of Symmetrical Sulfones from Rongalite: Expansion to Cyclic Sulfones by Ring-Closure Metathesis, *Eur. J. Org. Chem.* **2005**, pp. 3581–3585 (2005).

374. Y.-Q. Li and L.-P. Zhang, A Simple and Efficient Direct Method for the Synthesis of Symmetric Dibenzyl Sulfones from Sodium Dithionite and Benzyl Chlorides in Ionic Liquid, *Eur. J. Org. Chem.* **137**, pp. 1315–1319 (2006).

375. J. Bremner, M. Julia, M. Launay and J.-P. Stacino, Syntheses a Laide de Sulfones (xxiv). Synthese Stereoselective D'olefines par Hydrogenolyse des Vinylsulfones, *Tetrahedron Lett.* **23**, pp. 3265–3266 (1982).

376. M. Julia, H. Lauron, J.-P. Stacino and J.-N. Verpeaux, Organic Synthesis with Sulfones xxxviii. On the Mechanism of the Stereospecific Hydrogenolysis of Vinylic Sulfones by Sodium Dithionite, *Tetrahedron* **42**, pp. 2475–2484 (1986).

377. M. Julia and J.-M. Paris, Synthesis a Laide de Sulfones $v^{(+)}$- Methode de Synthese Generale de Doubles Liaisons, *Tetrahedron Lett.* **14**, pp. 4833–4836 (1973).

378. P. R. Blakemore, The Modified Julia Olefination: Alkene Synthesis via the Condensation of Metallated Heteroarylalkylsulfones with Carbonyl Compounds, *J. Chem. Soc. Perkin Trans.* **1**, pp. 2563–2585 (2002).

379. K. K. Park, C. W. Lee and S. Y. Choi, Viologen-Mediated Reductive Desulfonylation of α-nitro Sulfones by Sodium Dithionite, *J. Chem. Soc. Perkin Trans.* **1**, pp. 601–603 (1992).

380. R. Borgogno, S. Colonna and R. Fornasier, Reduction of Organic Sulfur Compounds by Formamidinesulfinic Acid Under Phase-Transfer Conditions, *Synthesis*, **7**, pp. 529–531 (1975).

381. J. Drabowicz and M. Mikolajczyk, An Iodine-Catalyzed Reduction of Sulphoxides by Formamidinesulphinic Acid, *Synthesis*, **10**, p. 542 (1978).

382. J.-H. Liu, A.-T. Wu, M.-H. Huang, C.-W. Wu and W.-S. Chung, The Syntheses of Pyrazino-Containing Sultines and their Application in Diels–Alder Reactions with Electron-Poor Olefins and [60] Fullerene, *J. Org. Chem.* **65**, pp. 3395–3403 (2000).

383. L. Tschugaeff and W. Chlopin, Beitrage zur Kenntnis des Reduktionsvermögens der Schwefligen Säure. i. Einwirkung von Natriumhydrosulfit auf Tellur und Selen, *Chem. Ber.* **47**, pp. 1269–1275 (1914).

384. M. P. Balfe, C. A. Chaplin and H. Phillips, The Oxidation of Certain Alkyl Tellurides, *J. Chem. Soc.*, pp. 341–347 (1938).

385. J. D. McCullough, The Synthesis of 1-thia-4-Telluracyclohexane (1,4-thiatellurane) and Four of its Addition Compounds, *Inorg. Chem.* **4**, pp. 862–864 (1965).

386. W. Lohner and K. Praefcke, Eine Verbesserte Synthese von Tellurophen, *Chem. Ber.* **111**, pp. 3745–3746 (1978).

387. H. Suzuki and M. Inouye, A Mild and Efficient Debromination of Vicinal Dibromoalkanes with Sodium Telluride Prepared from Tellurium and Rongalite, *Chem. Lett.* **14**, pp. 225–228 (1985).

388. H. Suzuki, H. Manabe and M. Inouye, Reduction of Aromatic Nitro Compounds with Sodium Telluride, *Chem. Lett.* **14**, pp. 1671–1674 (1985).

389. G. Polson and D. C. Dittmer, Functional Group Modification via Organotellurium Chemistry. Synthesis of Alkyl Alcohols from Chloromethyloxiranes, *Tetrahedron Lett.* **27**, pp. 5579–5582 (1986).

390. G. Discordia and D. C. Dittmer, 2-substituted-4-Hydroxymethyltellurophenes from Acetylenic Epichlorohydrides and Sodium Telluride, *Tetrahedron Lett.* **29**, pp. 4923–4926 (1988).

391. T. Junk, G. Gritzner, and K. J. Irgolic, Improved Preparation of Tetrahydroselenophene (Selenacyclopentane) and Tetrahydrotellurophene (Telluracyclopentane), *Synth. React. Inorg. Met.-Org. Chem.* **19**, pp. 931–936 (1989).

392. R. P. Discordia and D. C. Dittmer, Stereospecific Telluride-Mediated Conversion of Glycidols to Allyl Alcohols: An Extension of the Sharpless Kinetic Resolution, *J. Org. Chem.* **55**, pp. 1414–1415 (1990).

393. D. C. Dittmer, R. P. Discordia, Y. Zhang, C. K. Murphy, A. Kumar, A. S. Pepito and Y. Wang, A Tellurium Transposition Route to Allylic Alcohols: Overcoming Some Limitations of the Sharpless-Katsuki Asymmetric Epoxidation, *J. Org. Chem.* **58**, pp. 718–731 (1993).

394. A. Kumar and D. C. Dittmer, A Catalytic Process for the Transposition of Allylic Hydroxy Groups and Carbon–Carbon Double Bonds, *Tetrahedron Lett.* **35**, pp. 5583–5586 (1994).

395. Q. Xu, B. Chao, Y. Wang and D. C. Dittmer, Tellurium in the "No-solvent" Organic Synthesis of Allylic Alcohols, *Tetrahedron* **53**, pp. 12131–12146 (1997).

396. B. Chao and D. C. Dittmer, A Tellurium-Triggered Domino Reaction for the Synthesis of a 1-Substituted-3-vinyl-1,3-Dihydroisobenzofuran, *Tetrahedron Lett.* **41**, pp. 6001–6004 (2000).

397. S. Diaz, J. Cuesta, A. González and J. Bonjoch, Synthesis of (—)-Nakamurol A and Assignment of Absolute Configuration of Diterpenoid (+)-Nakamurol A, *J. Org. Chem.* **68**, pp. 7400–7406 (2000).

398. W. Guo, J. Chen, D. Wu, J. Ding, F. Chen and H. Wu, Rongalite® Promoted Highly Regioselective Synthesis of β-Hydroxy Sulfides by Ring Opening of Epoxides with Disulfides, *Tetrahedron* **65**, pp. 5240–5243 (2009).

399. W. Guo, G. Lv, J. Chen, W. Gao, J. Ding and H. Wu, Rongalite® and Base-Promoted Cleavage of Disulfides and Subsequent Michael Addition to α,β-Unsaturated Ketones/Esters: An Odorless Synthesis of β-Sulfido Carbonyl Compounds, *Tetrahedron* **66**, pp. 2297–2300 (2010).

400. V. Ganesh and S. Chandrasekaran, One-Pot Synthesis of β-Amino/β-Hydroxy Selenides and Sulfides from Aziridines and Epoxides, *Synthesis* **19**, pp. 3267–3278 (2009).

401. Z.-L. Wang, R.-Y. Tang, P.-S. Luo, C.-L. Deng, P. Zhong and J.-H. Li, Hydrothiolation of Terminal Alkynes with Diaryl Disulfides and Diphenyl Diselenide: Selective Synthesis of (z)-1-Alkenyl Sulfides and Selenides, *Tetrahedron* **64**, pp. 10670–10675 (2008).

402. W. Dan, H. Deng, J. Chen, M. Liu, J. Ding and H. Wu, A New Odorless One-Pot Synthesis of Thioesters and Selenoesters Promoted by Rongalite®, *Tetrahedron* **66**, pp. 7384–7388 (2010).

403. D. Crich and Y. Zhou, Catalytic Allylic Oxidation with a Recyclable, Fluorous Seleninic Acid, *Org. Lett.* **6**, pp. 775–777 (2004).

404. R. Golmohammadi, Studies on the Methods of Preparation of Di- and Tri-Selena-Straight-Chain Fatty Acids and the Corresponding Amides, with Some Remarks on the Infrared Spectra of the Mono-, Di- and Tri-selena-Fatty Acids, *Acta Chem. Scand.* **20**, pp. 563–571 (1966).

405. H. J. Reich, F. Chow and S. K. Shah, Selenuim Stabilized Carbanions. Preparation of α-Lithio Selenides and Applications to the Synthesis of Olefins by Reductive Elimination of β-Hydroxy Selenides and Selenoxide Syn Elimination, *J. Amer. Chem. Soc.*, **101**, pp. 6638–6648 (1979).

406. N. Geoffroy and G. P. Demopoulos, Reductive Precipitation of Elemental Selenium from Selenious Acidic Solutions Using Sodium Dithionite, *Ind. Eng. Chem. Res.* **48**, pp. 239–244 (2009).

407. J. T. B. Ferreira, A. R. M. de Oliveira and J. V. Comasseto, A Convenient Method of Synthesis of Dialkyltellurides and Dialkylditellurides, *Synth. Commun.* **19**, pp. 239–244 (1989).

408. J. V. Comasseto, E. S. Lang, J. T. B. Ferreira, F. Simonelli and V. R. Correia, The Use of Aminoiminomethanesulfinic Acid (Thiourea Dioxide) under Phase Transfer Conditions for Generating Organochalcogenate Anions. Synthesis of Sulfides, Selenides and Tellurides, *J. Organomet. Chem.* **334**, pp. 329–340 (1987).

409. E. S. Lang and J. V. Comasseto, Reduction of Organoselenium and Tellurium Halides and Oxides with Thiourea Dioxide, *Synth. Commun.* **18**, pp. 301–305 (1988).

410. J. G. de Vries, T. J. van Bergen and R. M. Kellogg, Sodium Dithionite as a Reductant for Aldehydes and Ketones, *Synthesis* **9**, pp. 246–247 (1977).

411. H. Minato, S. Fujie, K. Okuma and M. Kobayashi, Reduction of Pyridinium and Carbonyl Compounds by Refluxing Sodium Dithionite Solutions, *Chem. Lett.* **6**, pp. 1091–1094 (1977).

412. T. M. Olson, S. D. Boyce and M. R. Hoffmann, Kinetics, Thermodynamics, and Mechanism of the Formation of Benzaldehyde-S(IV) Adducts, *J. Phys. Chem.* **90**, pp. 2482–2488 (1986).

413. K. Ageishi, T. Endo and M. Okawara, Reduction of Aldehydes and Ketones by Sodium Dithionite Using Viologens as Electron Transfer Catalyst, *J. Polymer. Sci.: Polymer Chem. Ed.* **21**, pp. 175–181 (1983).

414. S. Chandrasekhar and M. Srimannarayana, New Approaches to the Cannizzaro and Tishchenko Reactions, *Synth. Commun.* **39**, pp. 4473–4478 (2009).

415. S. V. Makarov, E. V. Kudrik, R. van Eldik and E. V. Naidenko, Reactions of Methyl Viologen and Nitrite with Thiourea Dioxide. New Opportunities for an Old Reductant, *J. Chem. Soc. Dalton Trans.* **31** pp. 4074–4076 (2002).

416. O. Louis-Andre and G. Gelbard, Exclusive 1-4 Reduction of Conjugated Ketones by Sodium Dithionite, *Tetrahedron Lett.* **26**, pp. 831–832 (1985).

417. R. S. Dhillon, R. P. Singh and D. Kaur, Selective 1,4-Reduction of Conjugated Aldehydes and Ketones in the Presence of Unconjugated Aldehydes and Ketones with Sodium Dithionite, *Tetrahedron Lett.* **36**, pp. 1107–1108 (1995).

418. J. Singh, G. L. Kad, M. Sharma and R. S. Dhillon, Chemoselective Carbonyl Reduction of Functionalized Aldehydes and Ketones to Alcohols with Sodium Dithionite, *Synth. Commun.* **28**, pp. 2253–2257 (1998).

419. F. Camps, J. Coll and M. Riba, Stereochemistry of Methylcyclohexanones Reduction by Sodium Dithionite under Conventional and Phase Transfer Conditions, *J. Chem. Soc. Chem. Commun.* **15**, pp. 1080–1081 (1979).

420. F. Camps, J. Coll, A. Guerrero, J. Guitart and M. Riba, Reduction of Conjugated Dienoic Carboxylic Acids and Esters with Sodium Dithionite, *Chem. Lett.* **11**, pp. 715–718 (1982).

421. F. Camps, J. Coll and J. Guitart, Reduction of Unsaturated Conjugated Ketones with Sodium Dithionite under Phase Transfer Catalysis, *Tetrahedron* **42**, pp. 4603–4609 (1986).

422. F. Camps, J. Coll and J. Guitart, Reduction of Alkyl 2,4-Alkadienoates with Sodium Dithionite under Phase Transfer Conditions, *Tetrahedron* **43**, pp. 2329–2334 (1987).

423. K. G. Akamanchi, H. C. Patel and R. Meenakshi, Reduction of Steroidal Ketones by Sodium Dithionite in Presence of Phase Transfer Catalyst, *Synth. Commun.* **22**, pp. 1655–1660 (1992).

424. T. van Es and O. G. Backeberg, A New Synthesis of para-Substituted Benzils, and Preparation of Some of the Corresponding Benzoins, *J. Chem. Soc.*, pp. 1371–1373 (1963).

425. S. M. Heilmann, J. K. Rasmussen and H. K. Smith II, Reduction of Unsymmetrical Benzils Using Sodium Dithionite, *J. Org. Chem.* **48**, pp. 987–992 (1983).

426. A. R. Harris and T. J. Mason, The Reduction of Aromatic Aldehydes and Benzils by Sodium Formaldehyde Sulphoxylate, *Synth. Commun.* **19**, pp. 529–535 (1989).

427. A. R. Harris, Sodium Formaldehyde Sulphoxylate an Efficient Reagent for the Reduction of α-Haloketones, *Synth. Commun.* **17**, pp. 1587–1592 (1987).

428. W. F. Jarvis, M. D. Hoey, A. L. Finocchio and D. C. Dittmer, Sodium Formaldehyde Sulphoxylate an Efficient Reagent for the Reduction of α-Haloketones, *J. Org. Chem.* **53**, pp. 5750–5776 (1988).

429. K. Nakagawa and K. Minami, Reduction of Organic Compounds with Thiourea Dioxide. I. Reduction of Ketones to Secondary Alcohols, *Tetrahedron Lett.* **13**, pp. 343–346 (1972).

430. S.-L. Huang and T.-Y. Chen, Reduction of Organic Compounds with Thiourea Dioxide. I. Reduction of Aldehydes to Primary Alcohols, *J. Chin. Chem. Soc.* **21**, pp. 235–241 (1974).

431. R. B. dos Santos, T. J. Brocksom and U. Brocksom, A Convenient Deoxygenation of α,β-Epoxy Ketones to Enones, *Tetrahedron Lett.* **38**, pp. 745–748 (1997).

432. S. Sambher, C. Baskar and R. S. Dhillon, Novel Chemoselective Reduction of Aldehydes in the Presence of Other Carbonyl Moieties with Thiourea Dioxide, *Synth. Commun.* **38**, pp. 2150–2157 (2008).

433. S. Sambher, C. Baskar and R. S. Dhillon, Chemoselective Reduction of Carbonyl Groups of Aromatic Nitro Carbonyl Compounds to the Corresponding Nitroalcohols Using Thiourea Dioxide, *Arkivoc* **x**, pp. 141–145 (2009).

434. S. Urvashi and R. S. Dhillon, Microwave-Assisted Efficient and Chemoselective Reduction of Aldehydes with Thiourea Dioxide, *Asian J. Res. Chem.* **6**, pp. 158–162 (2013).

435. J. E. Herz and L. A. de Márquez, Reduction of Steroidal Ketones with Aminoiminomethanesulphinic Acid, *J. Chem. Soc. Perkin Trans. I* pp. 2633–2634 (1973).

436. R. Caputo, L. Mangoni, P. Monaco, G. Palumbo and L. Previtera, The Role of Thiourea S,S-Dioxide in the Reduction of Steroidal Ketones, *Tetrahedron Lett.* **16**, pp. 1041–1042 (1975).

437. N. Chatterjie, C. E. Inturrisi, H. B. Dayton and H. Blumberg, Stereospecific Synthesis of the 6β-Hydroxy Metabolites of Naltrexone and Naloxone, *J. Med. Chem.* **18**, pp. 490–492 (1975).

438. N. Chatterjie, J. G. Umans and C. E. Inturrisi, Reduction of 6-Ketones of the Morphine Series with Formamidinesulfinic Acid. Stereoselectivity Opposite to That of Hydride Reductions, *J. Org. Chem.* **41**, pp. 3624–3625 (1976).

439. P. H. Gore, Thiourea Dioxide as an Organic Reducing Agent, *Chem. Ind.* p. 1355 (1954).

440. S.-L. Huang and T.-Y. Chen, Reduction of Organic Compounds with Thiourea Dioxide II. The Reduction of Organic Nitrogen Compounds, *J. Chin. Chem. Soc.* **22**, pp. 91–94 (1975).

441. K. Nakagawa, S. Mineo, S. Kawamura and K. Minami, Reduction of Organic Compounds with Thiourea Dioxide. II. Reduction of Aromatic Nitro Compounds and Synthesis of Hydrazo Compounds, *Yakugaku Zasshi* **97**, pp. 1253–1256 (1977).

442. S. V. Makarov, E. V. Kudrik and K. A. Davydov, Reaction of Thiourea S,S-Dioxides with Dyes Containing Carbonyl or Azo Groups, *Russ. J. Gen. Chem.* **76**, pp. 1599–1603 (2006).

443. N. Chatterjie, A. Minar and D. D. Clarke, Synthesis of 2-Aminomorphine and 2-Aminocodeine. Reduction of Aromatic Nitro Groups with Formamidinesulphinic Acid, *Synth. Commun.* **9**, pp. 647–657 (1979).

444. J. Rosewear and J. F. K. Wilshire, The Reduction of Some o-Nitrophenylazo Dyes with Thiourea S,S-Dioxide (Formamidinesulfinic Acid): A General Synthesis of 2-Aryl-2H-benzotriazoles and Their 1-Oxides, *Austr. J. Chem.* **37**, pp. 2489–2497 (1984).

445. J. Rosewear and J. F. K. Wilshire, Preparation of Some 2-(Methoxyphenyl)-2H-benzotriazoles and the Corresponding Hydroxyphenyl Compounds, *Austr. J. Chem.* **40**, pp. 1663–1673 (1987).

446. S. Tanimoto and T. Kamano, Preparation of 2-(2H-Benzotriazol-2-yl)phenols by the Reduction of 2-[(2-Nitrophenyl)azo]phenols with Thiourea S,S-Dioxide, *Synthesis* **18**, pp. 647–649 (1986).

447. J. F. K. Wilshire, The Reduction of Some 2,2-Dinitrodiaryl Compounds and Related Compounds by Thiourea S,S-Dioxide (Formamidinesulfinic Acid), *Austr. J. Chem.* **41**, pp. 995–1001 (1988).

448. W. E. Truce, E. M. Kreider and W. W. Brand, The Smiles and Related Rearrangements of Aromatic Systems, *Org. React.* **18**, pp. 99–215 (1970).

449. I. N. Sokolova, V. V. Budanov, Y. V. Polenov and I. P. Polyakova, Kinetics of Reduction of 2-Nitro-2-oxy-5-methylazobenzene with Rongalite, *Izv. Vuzov Khim. Khim. Tekhnol.* **26**, pp. 822–825 (1983).

450. B. Quiclet-Sire, I. Thevenot and S. Z. Zard, A New and Practical Synthesis of Pyrroles, *Tetrahedron Lett.* **36**, pp. 9469–9470 (1995).

451. A. Kreutzberger and U. H. Tesch, Cyclosierungsversuche an Aminoimonomethansulfinsäure mit β-Diketonen, *Arch. Pharm.* **311**, pp. 429–432 (1978).

452. R. Balicki and U. Cmielowiec, A Mild Deoxygenation of Heteroaromatic N-Oxides by Formamidinesulfinic Acid, *Monatsh. Chem.* **131**, pp. 1105–1107 (2000).

453. W. B. Koniecky and F. A. L. Linch, Determination of Aromatic Nitro Compounds, *Anal. Chem.* **30**, pp. 1134–1137 (1958).

454. S. I. Obtemperanskaya and V. K. Zlobin, Application of Formamidinesulfinic Acid for Separate Determination of p-, o-, m-Nitrophenols, *Zh. Anal. Khim.* **29**, pp. 609–611 (1974).

455. S. I. Obtemperanskaya and V. K. Zlobin, Application of Thiourea Dioxide in Organic Analysis. Determination of Ethers of Nitric Acid, Nitroso- and Azocompounds, *Vestn. MGU. Khimia* **15**, pp. 247–249 (1974).

456. M. M. Buzlanova, I. V. Karandi and S. I. Obtemperanskaya, Photometric Determination of Nitrostilbenes Using Formamidinesulfinic Acid, *J. Anal. Chem.* **54**, pp. 1010–1012 (1999).

457. W. E. Curtis, M. E. Muldrow, N. B. Parker, R. Barkley, S. L. Linas and J. E. Repine, N,N′-Dimethylthiourea Dioxide Formation from N,N-Dimethylthiourea Reflects Hydrogen Peroxide Concentrations in Simple Biological Systems, *Proc. Natl. Acad. Sci. USA* **85**, pp. 3422–3425 (1988).

458. P. Roy and A. Pramanik, One-pot Sequential Synthesis of 1,2-Disubstituted Benzimidazoles under Metal-free Conditions, *Tetrahedron Lett.* **54**, pp. 5243–5245 (2013).

459. R. A. Scheuerman and D. Tumelty, The Reduction of Aromatic Nitro Groups on Solid Supports Using Sodium Hydrosulfite ($Na_2S_2O_4$), *Tetrahedron Lett.* **41**, pp. 6531–6535 (2000).

460. R. Kaplánek and V. Krchnak, Fast and Effective Reduction of Nitroarenes by Sodium Dithionite under PTC Conditions: Application in Solid-Phase Synthesis, *Tetrahedron Lett.* **54**, pp. 2600–2603 (2013).

461. K. K. Park, C. H. Oh and W. K. Joung, Sodium Dithionite Reduction of Nitroarenes Using Viologen as an Electron Phase-Transfer Catalyst, *Tetrahedron Lett.* **34**, pp. 7445–7446 (1993).

462. K. K. Park, C. H. Oh and W.-J. Sim, Chemoselective Reduction of Nitroarenes and Nitroalkanes by Sodium Dithionite Using Octylviologen as an Electron Transfer Catalyst, *J. Org. Chem.* **60**, pp. 6202–6204 (1995).

463. J. G. Carey, J. F. Cairns and J. E. Colchester, Reduction of 1,1′-Dimethyl-4,4′-Bipyridinium Dichloride to 1,1′-Dimethyl-1,1′-Dihydro-4,4′-Bipyridyl, *J. Chem. Soc. D: Chem. Commun.* **5**, pp. 1280–1281 (1969).

464. J. K. Kristjansson and T. C. Hollocher, First Practical Assay for Soluble Nitrous Oxide Reductase of Denitrifying Bacteria and a Partial Kinetic Characterization, *J. Biol. Chem.* **255**, pp. 704–707 (1980).

465. K. Tsukahara and R. G. Wilkins, Kinetics of Reduction of Eight Viologens by Dithionite Ion, *J. Am. Chem. Soc.* **107**, pp. 2632–2635 (1985).

466. D. H. P. Thompson, W. C. J. Barrette and J. K. Hurst, One-Electron Reduction of Dihexadecyl Phosphate Vesicle Bound Viologens by Dithionite Ion, *J. Am. Chem. Soc.* **109**, pp. 2003–2009 (1987).

467. H.-P. Kim, B. Claude and C. Tondre, Microenvironment Effects on the Kinetics of Electron-Transfer Reactions Involving Dithionite Ions and Viologens. 1. A Comparison between Two Types of Polyelectrolytes, *J. Phys. Chem.* **94**, pp. 7711–7716 (1990).

468. T. M. Bockman and J. K. Kochi, Isolation and Oxidation-Reduction of Methylviologen Cation Radicals. Novel Disproportionation in Charge-Transfer Salts by X-ray Crystallography, *J. Org. Chem.* **55**, pp. 4127–4135 (1990).

469. M. B. Yarmolinsky and S. P. Colowick, On the Mechanism of Pyridine Nucleotide Reduction by Dithionite, *Biochim. Biophys. Acta* **20**, pp. 177–189 (1956).

470. E. M. Kosower and S. W. Bauer, Pyridinium Complexes. II The Nature of Intermediate in the Dithionite Reduction of Diphosphopyridine Nucleotide DPN, *J. Am. Chem. Soc.* **82**, pp. 2191–2194 (1960).

471. W. S. Cauchey and K. A. Schellenberg, Characterization of an Intermediate in the Dithionite Reduction of a Diphosphopyridine Nucleotide Model as a 1,4-Addition Product by Nuclear Magnetic Resonance Spectroscopy, *J. Org. Chem.* **31**, pp. 1978–1982 (1966).

472. G. Blankenhorn and E. G. Moore, Sulfoxylate-ion (HSO_2^-), the Hydride Donor in Dithionite-dependent Reduction of NAD^+ Analogues, *J. Am. Chem. Soc.* **102**, pp. 1092–1098 (1980).

473. V. Carelli, F. Liberatore, L. Scipione, R. Musio and O. Sciaovelli, On the Strucure of Intermediate Adducts Arising from Dithionite Reduction of

Pyridinium Salts: A Novel Class of the Parent Sulfinic Acid, *Tetrahedron Lett.* **41**, pp. 1235–1240 (2000).

474. V. Carelli, F. Liberatore, L. Scipione, B. Di Rienzo and S. Tortorella, Dithionite Adducts of Pyridinium Salts: Regioselectivity of Formation and Mechanisms of Decomposition, *Tetrahedron* **61**, pp. 10331–10337 (2005).

475. R. Jokela, J. Miettinen and M. Lounasmaa, Sodium Dithionite Reduction of 1-[2-(3-Indolyl)Ethyl]Pyridinium Salts: Formation of a 1,2-Dihydropyridine Derivative via the Corresponding 1,4-Dihydropyridine Derivative, *Heterocycles* **32**, pp. 511–520 (1991).

476. C. E. Paul, I. W. C. E. Arends and F. Hollmann, Is Simpler Better? Synthetic Nicotinamide Cofactor Analogues for Redox Chemistry, *ACS Catal.* **4**, pp. 788–797 (2014).

477. S.-J. Lin and L. Guarente, Nicotinamide Adenine Dinucleotide, a Metabolic Regulator of Transcription, Longevity and Disease, *Curr. Opin. Cell Biol.* **15**, pp. 241–246 (2003).

478. A. A. Karyakin, Y. N. Ivanova and E. E. Karyakina, Equilibrium $(NAD^+/NADH)$ Potential on Poly (Neutral Red) Modified Electrode, *Electrochem. Commun.* **5**, pp. 677–680 (2003).

479. P. I. Dalko and L. Moisan, In the Golden Age of Organocatalysis, *Angew. Chem. Int. Ed.* **43**, pp. 5138–5175 (2004).

480. P. I. Dalko, *Enantioselective Organocatalysis: Reactions and Experimental Procedures*, 1st edn. Wiley-VCH Verlag GmbH & Co. KGaA, Weinheim (2007).

481. B. List, R. A. Lerner and C. F. Barbas III, Prolin-Catalyzed Direct Asymmetric Aldol Reactions, *J. Am. Chem. Soc.* **122**, pp. 2395–2396 (2000).

482. P. I. Dalko and L. Moisan, Enantioselective Organocatalysis, *Angew. Chem. Int. Ed.* **40**, pp. 3726–3748 (2001).

483. A. Berkessel and H. Gröger, *Asymmetric Organocatalysis*, 1st edn. Wiley-VCH Verlag GmbH & Co. KGaA, Weinheim (2005).

484. P. I. Dalko, *Comprehensive Enantioselective Organocatalysis: Catalysts, Reactions, and Applications*, 1st edn. Wiley-VCH Verlag GmbH & Co. KGaA, Weinheim (2013).

485. J.-A. Ma and D. Cahard, Towards Perfect Catalytic Asymmetric Synthesis: Dual Activation of the Electrophile and the Nucleophile, *Angew. Chem. Int. Ed.* **43**, pp. 4566–4583 (2004).

486. P. R. Schreiner, Metal-free Organocatalysis through Explicit Hydrogen Bonding Interactions, *Chem. Soc. Rev.* **32**, pp. 289–296 (2003).

487. V. Blažek-Bregović, N. Basarić and K. Mlinarić-Majerski, Anion Binding with Urea and Thiourea Derivatives, *Coord. Chem. Rev.* **295**, pp. 80–124 (2015).

488. M. S. Taylor and E. N. Jacobsen, Asymmetric Catalysis by Chiral Hydrogen-Bond Donors, *Angew. Chem. Int. Ed.* **45**, pp. 1520–1543 (2006).

489. A. G. Doyle and E. N. Jacobsen, Small-Molecule H-Bond Donors in Asymmetric Catalysis, *Chem. Rev.* **107**, pp. 5713–5743 (2007).

490. T. Okino, Y. Hoashi, T. Furukawa, X. Xu and Y. Takemoto, Enantio- and Diastereoselective Michael Reaction of 1,3-Dicarbonyl Compounds to

Nitroolefins Catalyzed by a Bifunctional Thiourea, *J. Am. Chem. Soc.* **127**, pp. 119–125 (2005).

491. Z. Zhang, Z. Bao and H. Xing, N,N'-Bis[3,5-bis(trifluoromethyl)-phenyl]thiourea: A Privileged Motif for Catalyst Development, *Org. Biomol. Chem.* **12**, pp. 3151–3162 (2012).

492. X. Fang and C.-J. Wang, Advances in Asymmetric Organocatalysis Mediated by Bifunctional Amine-Thioureas Bearing Multiple Hydrogen-Bonding Donors, *Chem. Commun.* **51**, pp. 1185–1197 (2015).

493. S. Verma and S. L. Jain, Thiourea Dioxide in Water as a Recyclable Catalyst for the Synthesis of Structurally Diverse Dihydropyrido[2,3-d]pyrimidine-2,4-diones, *Tetrahedron Lett.* **53**, pp. 2595–2600 (2012).

494. S. S. Mansoor, K. Logaiya, K. Aswin and P. N. Sudhan, An Appropriate One-Pot Synthesis of 3,4-Dihydropyranol[c]chromenes and 6-Amino-5-cyano-4-aryl-2-methyl-4H-pyrans with Thiourea Dioxide as an Efficient, Reusable Organic Catalyst in Aqueous Medium, *J. Taibah Uni. Sci.* **9**, pp. 213–226 (2015).

495. P. S. Bhale, S. B. Dongare and Y. B. Mule, An Efficient Synthesis of 1,8-Dioxooctahydroxanthenes Catalysed by Thiourea Dioxide (TUD) in Aqueous Media, *Chem. Sci. Trans.* **4**, pp. 246–250 (2015).

496. N. Kumar, S. Verma and S. L. Jain, An Efficient Reusable Catalyst for the Synthesis of Polyhydroquinolines via Hantzsch Multicomponent Coupling, *Chem. Lett.* **41**, pp. 920–922 (2012).

497. Z. Du and Z. Shao, Combining Transition Metal Catalysis and Organocatalysis — An Update, *Chem. Soc. Rev.* **42**, pp. 1337–1378 (2013).

498. H. Pellisier, *Enantioselective Multicatalyzed Tandem Reactions*, 1st edn. RSC Publishing, Cambridge (2014).

499. A. K. Geim and K. S. Novoselov, The Rise of Graphenes, *Nat. Mater.* **6**, pp. 183–191 (2007).

500. S. Pei and H.-M. Cheng, The Reduction of Graphene Oxide, *Carbon* **50**, pp. 3210–3228 (2012).

501. D. R. Dreyer, S. Park, C. W. Bielawski and R. S. Ruoff, The Chemistry of Graphene Oxide, *Chem. Soc. Rev.* **39**, pp. 228–240 (2010).

502. M. Wissler, Graphite and Carbon Powders for Electrochemical Applications, *J. Power Sources* **156**, pp. 142–150 (2006).

503. T. Kuila, A. K. Mishra, P. Khanra, N. H. Kim and J. H. Lee, Recent Advances in the Efficient Reduction of Graphene Oxide and Its Application as Energy Storage Electrode Materials, *Nanoscale* **5**, pp. 52–71 (2013).

504. T. Zhou, F. Chen, K. Liu, H. Deng, Q. Zhang, J. Feng and Q. Fu, A Simple and Efficient Method to Prepare Graphene by Reduction of Graphite Oxide with Sodium Hydrosulfite, *Nanotechnology* **22**, pp. 045704–045710 (2011).

505. L. Sun, H. Yu and B. Fugetsu, Graphene Oxide Adsorption Enhanced by *in situ* Reduction with Sodium Hydrosulfite to Remove Acridine Orange from Aqueous Solution, *J. Hazard. Mater.* **203–204**, pp. 101–110 (2012).

506. J. Molina, J. Fernandez, A. I. del Rio, J. Bonastre and F. Cases, Chemical and Electrochemical Study of Fabrics Coated with Reduced Graphene Oxide, *Appl. Surf. Sci.* **279**, pp. 46–54 (2013).

507. L.-B. Xing, J.-L. Zhang, J. Zhang, S.-F. Hou, J. Zhou, W. Si, H. Cui and S. Zhuo, Three Dimensional Reduced Graphene Hydrogels with Tunable Pore Sizes Using Thiourea Dioxide for Electrode Materials in Supercapacitors, *Electrochim. Acta* **176**, pp. 1288–1295 (2015).

508. R. G. Wilkins, *Kinetics and Mechanisms of Reactions of Transition Metal Complexes*, 1st edn. VCH Publications, Weinheim (1991).

509. P. A. Loach and M. Calvin, Oxidation States of Manganese Hematoporphyrin IX in Aqueous Solution, *Biochemistry* **2**, pp. 361–371 (1963).

510. J. James and P. Hambright, Kinetics of the Oxidation of Dithionite $(S_2O_4^{2-})$ by Manganese (III) — Hematoporphyrin, *J. Coord. Chem.* **3**, pp. 183–185 (1973).

511. P. Hambright and P. B. Chock, Metal-Porphyrin Interactions. IV. Electron-Transfer Kinetics between Dithionite and Manganese (III) and Cobalt (III) Porphyrins, *Inorg. Chem.* **13**, pp. 3029–3031 (1974).

512. P. Worthington and P. Hambright, Kinetics of the Oxidation of Dithionite by Dicianoporphyrinato Ferrate (III) Complexes. *J. Inorg. Nucl. Chem.* **42**, pp. 1651–1654 (1980).

513. A. Adeyemo, A. Valiotti and P. Hambright, Kinetics of the Reduction of Cyano Cobalt (III) Porphyrins by Dithionite, *Inorg. Chim. Acta Lett.* **64**, pp. L251–L253 (1982).

514. C. W. J. Scaife and R. G. Wilkins, Kinetics of the Reduction of Hexacyanoferrate (III) by Dithionite Ion. *Inorg. Chem.* **19**, pp. 3244–3247 (1980).

515. P. Hambright, S. Lemelle, K. Alston, P. Neta, H. H. Newball and S. Di Stefano, A Dissociative Mechanism for the Dithionite Reduction of Cobalt (III) Myoglobin, *Inorg. Chim. Acta* **92**, pp. 167–172 (1984).

516. E. Olivas, D. J. A. de Waal and R. G. Wilkins, Reduction of Metmyoglobin Derivatives by Dithionite Ion, *J. Biol. Chem.* **252**, pp. 4038–4042 (1977).

517. P. C. Harrington, D. J. A. de Waal and R. G. Wilkins, Reduction of Methemerythrin by Dithionite Ion, *Arch. Biochem. Biophys.* **191**, pp. 444–451 (1978).

518. D. R. Eaton and R. G. Wilkins, Reduction by Dithionite Ion of Adducts of Metmyoglobin with Imidazole, Pyridine, and Derivatives, *J. Biol. Chem.* **253**, pp. 908–915 (1978).

519. Z. Bradić, K. Tsukahara, P. C. Wilkins and R. G. Wilkins, Reduction by Radicals of Metmyoglobin, Methemerythrin and Derivatives, *Rev. Port. Quim.* **27**, pp. 128–129 (1985).

520. K. Tsukahara and K. Ishida, Kinetics and Mechanism of Reduction of Metmyoglobins by Dithionite. Role of the Heme Propionates, *Bull. Chem. Soc. Jpn.* **64**, pp. 2378–2382 (1991).

521. R. N. Mehrotra and R. G. Wilkins, Kinetics of Reduction of Metal Complexes by Dithionite, *Inorg. Chem.* **19**, pp. 2177–2178 (1980).

522. Z. Bradić and R. G. Wilkins, Comparative Behavior in the Kinetics of Reduction by Superoxide and Dithionite Ions, *J. Am. Chem. Soc.* **106**, pp. 2236–2239 (1984).

523. R. Langley, P. Hambright, K. Alston and P. Neta, Kinetics of the Reduction of Manganese (III) Myoglobin by Dithionite, *Inorg. Chem.* **25**, pp. 114–117 (1986).

524. S. Lynn, R. G. Rinker and W. H. Corcoran, The Monomerization Rate of Dithionite Ion in Aqueous Solution, *J. Phys. Chem.* **68**, p. 2363 (1964).

525. D. Pinnell and R. B. Jordan, Kinetics of Reduction of Cobalt (III)-Ammine Complexes by Dithionite, *Inorg. Chem.* **18**, pp. 3191–3194 (1979).

526. R. F. Pasternack, M. A. Cobb and N. Sutin, Substitution and Oxidation-Reduction Reactions of a Water-Soluble Porphyrin Tetrakis(4-N-methylpyridyl)porphinecobalt(III)-Pyridine System, *Inorg. Chem.* **14**, pp. 865–873 (1975).

527. M. J. Hintz and J. A. Peterson, The Kinetics of Reduction of Cytochrome P-450$_{cam}$ by the Dithionite Anion Monomer, *J. Biol. Chem.* **255**, pp. 7317–7325 (1980).

528. P.-I. Ohlsson, J. Blanck and K. Ruckpaul, Reduction of Lactoperoxidase by the Dithionite Anion Monomer, *Eur. J. Biolchem.* **158**, pp. 451–454 (1986).

529. J. Peterson and M. T. Wilson, The Reduction of Haem Peptides by Dithionite. A Kinetic Investigation, *Inorg. Chim. Acta* **135**, pp. 101–107 (1987).

530. J. C. W. Chien and L. C. Dickinson, Reduction of Cobalticytochrome c by Dithionite, *J. Biol. Chem.* **253**, pp. 6965–6972 (1978).

531. A. Haim, Comments on "Reduction of Cobalticytochrome c by Dithionite", *J. Phys. Chem.* **83**, p. 2553 (1979).

532. J. C. Cassatt, M. Kukuruzinska and J. W. Bender, Kinetics of Reduction of Hemin and the Hemin Bis(pyridine) Complex by Dithionite, *Inorg. Chem.* **16**, pp. 3371–3372 (1977).

533. R. J. Balahura and M. D. Johnson, Outer-Sphere Dithionite Reductions of Metal Complexes, *Inorg. Chem.* **26**, pp. 3860–3863 (1987).

534. E. V. Kudrik, S. V. Makarov, A. Zahl and R. van Eldik, Kinetics and Mechanism of the Cobalt Phthalocyanine Catalyzed Reduction of Nitrite and Nitrate by Dithionite in Aqueous Solution, *Inorg. Chem.* **42**, pp. 618–624 (2003).

535. E. V. Kudrik, S. V. Makarov, A. Zahl and R. van Eldik, Kinetics and Mechanism of the Iron Phthalocyanine Catalyzed Reduction of Nitrite by Dithionite and Sulfoxylate in Aqueous Solution, *Inorg. Chem.* **44**, pp. 6470–6475 (2005).

536. R. A. Larson and J. Cervini-Silva, Dechlorination of Substituted Trichloromethanes by an Iron (III) Porphyrin, *Environ. Toxicol. Chem.* **19**, pp. 543–548 (2000).

537. V. A. Nzengung, R. M. Castillo, W. P. Gates and G. L. Mills, Abiotic Transformation of Perchloroethylene in Homogeneous Dithionite Solution and in Suspensions of Dithionite-Treated Clay Minerals, *Environ. Sci. Technol.* **35**, pp. 2244–2251 (2001).

538. H. K. Boparai, P. I. Shea, S. D. Comfort and D. D. Snow, Dechlorinating Chloroacetanilide Herbicides by Dithionite-Treated Aquifer Sediment and Surface Soil, *Environ. Sci. Technol.* **40**, pp. 3043–3049 (2006).

539. A. Mot, Z. Kis, D. A. Svistunenko, G. Damian, R. Silaghi-Dumitrescu and S. V. Makarov, "Super-Reduced" Iron under Physiologically-Relevant Conditions, *Dalton Trans.* **39**, pp. 1464–1466 (2010).

540. S. V. Makarov, D. S. Salnikov, A. S. Pogorelova, Z. Kis and R. Silaghi-Dumitrescu, A New Route to Carbon Monoxide Adducts of Heme Proteins, *J. Porphyr. Phthalocya.* **12**, pp. 1096–1100 (2008).

541. A. S. Pogorelova, S. V. Makarov, E. S. Ageeva and R. Silaghi-Dumitrescu, Cobalt Phthalocyaninate as a Catalyst of the Reduction of Nitrite with Thiourea Dioxide, *Russ. J. Phys. Chem. A* **83**, pp. 2050–2053 (2009).

542. E. A. Vlasova, S. V. Makarov and M. N. Malinkina, The Kinetics of Nitrite Reduction by Thiourea Dioxide in the Presence of Cobalt Octasulfophenyltetrapyrazinoporphyrazine, *Russ. J. Phys. Chem. A* **84**, pp. 655–660 (2010).

543. E. A. Vlasova, A. S. Makarova, S. V. Makarov and E. S. Ageeva, Nitrite Reduction by Sodium Hydroxymethanesulfinate in the Presence of Cobalt or Iron Phthalocyanines, *Izv. Vyssh. Uchebn. Zaved. Khim. Khim Tekhnol.* **53**, pp. 74–78 (2010).

544. I. A. Dereven'kov, S. S. Ivanova, E. V. Kudrik, S. V. Makarov, A. S. Makarova and P. A. Stuzhin, Comparative Study of Reactions between μ-Nitrido- or μ-Oxo-bridged Iron Tetrasulfophthalocyanines and Sulfur-containing Reductants, *J. Serb. Chem. Soc.* **78**, pp. 1513–1530 (2013).

545. D. S. Salnikov, R. Silaghi-Dumitrescu, S. V. Makarov, R. van Eldik and G. R. Boss, Cobalamin Reduction by Dithionite. Evidence for the Formation of a Six-coordinate Cobalamin (II) Complex, *Dalton Trans.* **40**, pp. 9831–9834 (2011).

546. D. S. Salnikov, I. A. Dereven'kov, E. N. Artyushina and S. V. Makarov, Interaction of Cyanocobalamin with Sulfur-containing Reducing Agents in Aqueous Solutions, *Russ. J. Phys. Chem. A* **87**, pp. 44–48 (2013).

547. D. S. Salnikov, I. A. Dereven'kov, S. V. Makarov, E. S. Ageeva, A. Lupan, M. Surducan and R. Silaghi-Dumitrescu, Kinetics of Reduction of Cobalamin by Sulfoxylate in Aqueous Solutions, *Rev. Roum. Chim.* **57**, pp. 353–359 (2012).

548. I. A. Dereven'kov, D. S. Salnikov, S. V. Makarov, G. R. Boss and O. I. Koifman, Kinetics and Mechanism of Oxidation of Super-Reduced Cobalamin and Cobinamide Species by Thiosulfate, Sulfite and Dithionite, *Dalton Trans.* **42**, pp. 15307–15316 (2013).

549. I. A. Dereven'kov, D. S. Salnikov, R. Silaghi-Dumitrescu, S. V. Makarov and O. I. Koifman, Redox Chemistry of Cobalamin and Its Derivatives, *Coord. Chem. Rev.* **309**, pp. 68–83 (2016).

550. Q. Sun, A. J. Feitz, J. Guan and T. D. Waite, Comparison of the Reactivity of Nanosized Zero-Valent Iron (nZVI) Particles Produced by Borohydride and Dithionite Reduction of Iron Salts, *Nano* **3**, pp. 341–349 (2008).

551. X. Ma, D. He, A. M. Jones, R. N. Collins and T. D. Waite, Reductive Reactivity of Borohydride- and Dithionite-synthesized Iron-based Nanoparticles: A Comparative Study, *J. Hazard. Mater.* **303**, pp. 101–110 (2016).

552. Y. Xie and D. M. Cwiertny, Use of Dithionite to Extend the Reactive Lifetime of Nanoscale Zero-Valent Iron Treatment Systems, *Environ. Sci. Technol.* **44**, pp. 8649–8655 (2010).

553. E.-J. Kim, J. H. Kim, A.-M. Azad and Y.-S. Chang, Facile Synthesis and Characterization of Fe/FeS Nanoparticles for Environmental Applications, *ACS Appl. Mater. Interfaces* **3**, pp. 1457–1462 (2011).

554. R.-W. Gillham and S. F. O. O'Hannesin, Enhanced Degradation of Halogenated Aliphatics by Zero-Valent Iron, *Ground Water* **32**, pp. 958–967 (1994).

555. L. J. Matheson and P. G. Tratnyek, Reductive Dehalogenation of Chlorinated Methanes by Iron Metal, *Environ. Sci. Technol.* **28**, pp. 2045–2053 (1994).

556. C.-B. Wang and W.-X. Zhang, Synthesizing Nanoscale Iron Particles for Rapid and Complete Dechlorination of TCE and PCBs, *Environ. Sci. Technol.* **31**, pp. 2154–2156 (1997).

557. F. Fu, D. D. Dionysiou and H. Liu, The Use of Zero-Valent Iron for Groundwater Remediation and Wastewater Treatment: A Review, *J. Hazard. Mater.* **267**, pp. 194–205 (2014).

558. F.-D. Kopinke, G. Speichert, K. Mackenzie and E. Hey-Hawkins, Reductive Dechlorination in Water: Interplay of Sorption and Reactivity, *Appl. Catal. B: Environ.* **181**, pp. 747–753 (2016).

559. R. A. Crane and T. B. Scott, Nanoscale Zero-Valent Iron: Future Prospects for an Emerging Water Treatment Technology, *J. Hazard. Mater.* **211–212**, pp. 112–125 (2012).

560. D. O'Carroll, B. Sleep, M. Krol, H. Boparai and C. Kocur, Nanoscale Zero Valent Iron and Bimetallic Particles for Contaminated Site Remediation, *Adv. Water Resour.* **51**, pp. 104–122 (2013).

561. K. D. Grieger, A. Fjordborge, N. B. Hartmann, E. Eriksson, P. L. Bjerg and A. Baun, Environmental Benefits and Risks of Zero-valent Iron Nanoparticles (nZVI) for *In Situ* Remediation: Risk Mitigation or Trade-off? *J. Contam. Hydrol.* **118**, pp. 165–183 (2010).

562. A. B. Cundy, L. Hopkinson and R. L. D. Whitby, Use of Iron-Based Technologies in Contaminated Land and Groundwater Remediation: A Review, *Sci. Total Environ.* **400**, pp. 42–51 (2008).

563. E. Butler and K. F. Hayes, Effects of Solution Composition and pH on the Reductive Dechlorination of Hexachloroethane by Iron Sulfide, *Environ. Sci. Technol.* **32**, pp. 1276–1284 (1998).

564. E. Butler and K. F. Hayes, Kinetics of the Transformation of Halogenated Aliphatic Compounds by Iron Sulfide, *Environ. Sci. Technol.* **34**, pp. 422–429 (2000).

565. E. V. Egorova, S. V. Makarov, V. V. Budanov and D. N. Akbarov, Kinetics of Reduction of Ni^{2+} by Sodium Dithionite, *Zh. Obshch. Khim.* **61**, pp. 542–546 (1991).

566. P. K. Khanna, Reduction of Transition Metal Salts by SFS: Synthesis of Copper and Silver Sulphides, *Synth. React. Inorg. Met.-Org. Chem.* **37**, pp. 805–808 (2007).

567. P. K. Khanna, S. Gaikwad, P. V. Adhyapak, N. Singh and R. Marimuthu, Synthesis and Characterization of Copper Nanoparticles, *Mat. Lett.* **61**, pp. 4711–4714 (2007).

568. P. K. Khanna and V. V. V. S. Subbarao, Nanosized Silver Powder via Reduction of Silver Nitrate by Sodium Formaldehydesulfoxylate in Acidic pH Medium, *Mat. Lett.* **57**, pp. 2242–2245 (2003).

569. P. K. Khanna, N. Singh, D. Kulkarni, S. Deshmukh, S. Charan and P. V. Adhyapak, Water Based Simple Synthesis of Re-dispersible Silver Nanoparticles, *Mat. Lett.* **61**, pp. 3366–3370 (2007).

570. P. K. Khanna, T. S. Kale, M. Shaikh, N. K. Rao and C. W. Satyanarayanak, Synthesis of Oleic Acid Capped Copper Nano-Particles via Reduction of Copper Salt by SFS, *Mat. Chem. Phys.* **110**, pp. 21–25 (2008).

571. E. V. Egorova, S. V. Makarov, D. N. Akbarov and V. V. Budanov, Interaction of Sodium Hydroxymethanesulfinate with Nickel Salts. *Zh. Prikl. Khim.* **62**, pp. 972–976 (1989).

572. J. E. McGill and F. Lindstrom, Mechanism of Reduction of Cadmium by Aminoiminomethanesulfinic Acid in Alkaline Media, *Anal. Chem.* **49**, pp. 26–29 (1977).

573. S. V. Ermolina, S. V. Makarov, I. N. Terskaya and V. V. Budanov, Chemical Deposition of Nickel from Water and Water–Alcohol Solutions by Thiourea Dioxide, *Zh. Neorg. Khim.* **40**, pp. 1466–1469 (1995).

574. N. V. Sotskaya, S. V. Makarov, O. V. Dolgikh, V. M. Kashkarov, A. S. Lenshin and E. A. Kotlyarova, Surface Modification of Composites with Metal Nanoparticles, *Inorg. Mat.* **46**, pp. 1192–1197 (2010).

575. A. R. Fritzberg, D. M. Lyster and D. H. Dolphin, Evaluation of Formamidine Sulfinic Acid and Other Reducing Agents Agents for Use in the Preparation of Nc-99m Labeled Radiopharmaceuticals, *J. Nucl. Med.* **18**, pp. 553–557 (1977).

576. J. R. Scott, G. L. Garrett and B. C. Lentle, Preparation of Technetium-99m Glucoheptonate Utilizing Formamidine Sulfinic Acid, *Int. J. Nucl. Med. Biol.* **7**, pp. 71–73 (1980).

577. J. Baldas and P. M. Pojer, The Use of Formamidine Sulphinic Acid in the Preparation of ^{99m}Tc-Labelled Radiopharmaceuticals — A Cautionary Note, *Int. J. Nucl. Med. Biol.* **8**, pp. 110–111 (1981).

578. M. Neves, H. Ferronha and L. Patricio, Formamidine Sulfinic Acid as Reducing Agent in Technetium-99m Rhenium Sulfide Labelling, *J. Radioanal. Nucl. Chem.* **132**, pp. 241–249 (1989).

579. L. Selwyn and S. Tse, The Chemistry of Sodium Dithionite and Its Use in Conservation, *Stud. Conserv.* **53**(Supplement 2), pp. 61–73 (2008).

580. C. Pelé, E. Guilminot, S. Labroche, G. Lemoine and G. Baron, Removal from Waterlogged Wood: Extraction by Electrophoresis and Chemical Treatments, *Stud. Conserv.* **60**, pp. 155–171 (2015).

581. J. Torrent, U. Schwertmann and V. Barrón, The Reductive Dissolution of Synthetic Goethite and Hematite in Dithionite, *Clays Clay Miner.* **22**, pp. 329–337 (1987).

582. E. H. Rueda, M. C. Ballesteros and R. L. Grassi, Dithionite as a Dissolving Reagent for Goethite in the Presence of EDTA and Citrate. Application to Soil Analysis, *Clays Clay Miner.* **40**, pp. 575–585 (1992).

583. C. M. R. Abreu, P. V. Mendonca, A. C. Serra, A. V. Popov, K. Maty-jaszewski, T. Guliashvili and J. F. J. Coelho, Inorganic Sulfites: Efficient Reducing Agents and Supplemental Activators for Atom Transfer Radical Polymerization, *ACS Macro Lett.* **1**, pp. 1308–1311 (2012).

584. J. A. Davies, C. M. Hockensmith, V. Y. Kukushkin and Y. N. Kukushkin, *Synthetic Coordination Chemistry: Principles and Practice.* World Scientific, Singapore (1996).

585. F. Deac, N. Cotolan, Z. Kis and R. Silaghi-Dumitrescu, *A Dithionite-induced Six-coordinated Species at the Heme in Deoxy Hemoglobin, in Metal Elements in Environment, Medicine and Biology*, 1st edn. Publ. House Euro-bit, Timisoara, Romania (2010).

586. A. Das, R. Silaghi-Dumitrescu, L. G. Ljungdahl and D. M. Kurtz Jr, Cytochrome bd Oxidase, Oxidative Stress and Dioxygen Tolerance of the Strictly Anaerobic Bacterium Moorella thermoacetica, *J. Bacteriol.* **187**, pp. 2020–2029 (2005).

587. R. Quast, S. Paléus, E. Östlund and G. Bloom, Some Spectrophotometric Characteristics of Hemoglobin in the Atlantic Hagfish (Myxine glutinosa L.), *Acta Chem. Scand.* **21**, pp. 2849–2854 (1967).

588. J. M. Salhany, Effect of Carbon Dioxide on Human Hemoglobin, *J. Biol. Chem.* **247**, pp. 3799–3801 (1972).

589. R. P. Cox and M. R. Hollaway, The Reduction by Dithionite of Fe(III) Myoglobin Derivatives with Different Ligands Attached to the Iron Atom. A Study by Rapid-Wavelength-Scanning Stopped-Flow Spectrophotometry, *Eur. J. Biochem.* **74**, pp. 575–587 (1977).

590. J. M. Salhany, Kinetics of Reaction of Nitrite with Deoxy Hemoglobin after Rapid Deoxygenation or Predeoxygenation by Dithionite Measured in Solution and Bound to the Cytoplasmic Domain of Band 3 (SLC4A1), *Biochemistry* **47**, pp. 6059–6072 (2008).

591. R. M. Nalbandian, B. M. Nichols, F. R. Camp Jr, J. M. Lusher, N. F. Conte, R. L. Henry and P. L. Wolf, Dithionite Tube Test a Rapid, Inexpensive Technique for the Detection of Hemoglobin S and Non-S Sickling Hemoglobin, *Clin. Chem.* **17**, pp. 1028–1032 (1971).

592. J. H. Pajput and J. P. Naik, Optimization of Blood Sample Volume in Dithionite Tube Turbidity (DTT) Test with Respect to Hemoglobin Concentration Used for Field Screening of Sickle Hemoglobin (HbS), *Int. J. Pure Appl. Biosci.* **3**, pp. 325–330 (2015).

593. N. W. Tietz and E. A. Fiereck, The Spectrophotometric Measurement of Carboxyhemoglobin, *Ann. Clin. Lab. Sci.* **3**, pp. 36–42 (1973).

594. T. J. Siek and F. Rieders, Determination of Carboxyhemoglobin in the Presence of Other Blood Hemoglobin Pigments by Visible Spectrophotometry, *J. Forensic Sci.* **29**, pp. 39–54 (1984).

595. R. J. Lewis, R. D. Johnson and D. V. Canfield, An Accurate Method for the Determination of Carboxyhemoglobin in Postmortem Blood Using GC-TCD, *J. Anal. Toxicol.* **28**, pp. 59–62 (2004).

596. Y. Ohno, T. Kawanishi, A. Takahashi, Y. Kasuya and Y. Omori, A New Device for the Determination of Microsomal Cytochrome P-450 in Renal

Tissue Preparations from Various Species Contaminated with Mitochondria and Hemoglobin, *Jpn. J. Pharmacol.* **32**, pp. 679–688 (1982).

597. A. Dong, M. Nagai, Y. Yoneyama and W. S. Caughey, Determination of the Amounts and Oxidation States of Hemoglobins M Boston and M Saskatoon in Single Erythrocytes by Infrared Microspectroscopy, *J. Biol. Chem.* **269**, pp. 25365–25368 (1994).

598. S. M. Snell and M. A. Marini, A Convenient Spectroscopic Method for the Estimation of Hemoglobin Concentrations in Cell-free Solutions, *J. Biochem. Biophys. Meth.* **17**, pp. 25–34 (1988).

599. R. R. Borodulin, L. N. Kubrina, V. A. Serezhenkov, D. S. Burbaev, V. D. Mikoyan and A. F. Vanin, Redox Conversions of Dinitrosyl Iron Complexes with Natural Thiol-containing Ligands, *Nitric Oxide Biol. Chem.* **35**, pp. 35–41 (2013).

600. R. R. Borodulin, I. A. Dereven'kov, S. V. Makarov, V. D. Mikoyan, V. A. Serezhenkov, L. N. Kubrina, I. Ivanovic-Burmazovic and A. F. Vanin, Redox Activities of Mono- and Protein-bound Dinitrosyl Iron Complexes with Thiol-containing Ligands, *Nitric Oxide Biol. Chem.* **40**, pp. 100–109 (2014).

601. A. F. Vanin, *Dinitrosyl Iron Complexes with Thiols. Physicochemistry, Biology, Medicine*, 1st edn. Institute of Computer Studies, Moscow-Izhevsk (2015).

602. H. Lewandowska, J. Sadlo, S. Meczynska, T. M. Stepkowski, G. Wójciuk and M. Kruszewski, Formation of Glutathionyl Dinitrosyl Iron Complexes Protects Against Iron Genotoxicity, *Dalton Trans.* **44**, pp. 12640–12652 (2015).

603. R. J. Balahura, G. Ferguson, B. L. Ruhl and R. G. Wilkins, Kinetics of the Reduction of Nitro to Hydroxylamine Groups by Dithionite in a Cobalt(II) Cryptand Complex. X-ray Analysis of [1,8-Bis(hydroxylamino)-3,6,10,13,16,19-hexaazabicyclo[6.6.6]eicosane]cobalt(III) Chloride Tetrahydrate, *Inorg. Chem.* **22**, pp. 3990–3992 (1983).

604. P. A. Lay and A. M. Sargeson, Behavior of Nitro Substituents of Cage Complexes, *Inorg. Chem.* **29**, pp. 2762–2770 (1990).

605. H. H. Capps and W. M. Dehn, Desulfurization of Thioureas by Bromate and Iodate Solutions, *J. Am. Chem. Soc.* **54**, pp. 4301–4305 (1932).

606. M. Alamgir and I. R. Epstein, Complex Dynamical Behavior in a New Chemical Oscillator: The Chlorite-Thiourea Reaction in a CSTR, *Int. J. Chem. Kinet.* **17**, pp. 429–439 (1985).

607. M. Alamgir and I. R. Epstein, New Chlorite Oscillators: Chlorite-Bromide and Chlorite-Thiocyanate in a CSTR, *J. Phys. Chem.* **89**, pp. 3611–3614 (1985).

608. R. H. Simoyi, J. Masere, C. Muzimbaranda, M. Manyonda and S. Dube, Travelling Wave in the Chlorite-Thiourea Reaction, *Int. J. Chem. Kinet.* **23**, pp. 419–429 (1991).

609. I. R. Epstein, K. Kustin and R. H. Simoyi, Kinetics and Mechanism of the Chlorite-Thiourea Reaction in Acidic Medium, *J. Phys. Chem.* **96**, pp. 5852–5856 (1992).

610. G. Rábai, T. X. Wang and K. Kustin, Kinetics and Mechanism of the Oxidation of Thiourea by Chlorine Dioxide, *Int. J. Chem. Kinet.* **25**, pp. 53–62 (1993).

611. G. Rábai and M. Orbán, General Model for the Chlorite Ion Based Chemical Oscillators, *J. Phys. Chem.* **97**, pp. 5935–5939 (1993).

612. I. Lengyel, L. Györgyi and I. R. Epstein, Analysis of a Model of Chlorite-Based Chaotic Chemical Oscillators, *J. Phys. Chem.* **99**, pp. 12804–12808 (1995).

613. Q. Y. Gao, B. M. Li, K. Sun, Z.-S. Cai and X. Z. Zhao, Nonlinear Dynamics in ClO_2–$SC(NH_2)_2$ Reaction System, *Acta Phys.-Chim. Sin.* **17**, pp. 257–260 (2001).

614. Q. Gao and J. Wang, pH Oscillations and Complex Reaction Dynamics in the Non-buffered Chlorite–Thiourea Reaction, *Chem. Phys. Lett.* **391**, pp. 349–353 (2004).

615. J. B. Jones, C. R. Chinake and R. H. Simoyi, Oxyhalogen-Sulfur Chemistry: Oligooscillations in the Formamidinesulfinic Acid-Chlorite Reaction, *J. Phys. Chem.* **99**, pp. 1523–1529 (1995).

616. C. J. Doona and D. M. Stanbury, Adventitious Catalysis in Oscillatory Reductions by Thiourea, *J. Phys. Chem.* **98**, pp. 12630–12634 (1994).

617. T. R. Chigwada, E. Chikwana and R. H. Simoyi, S-Oxygenation of Thiocarbamides I: Oxidation of Phenylthiourea by Chlorite in Acidic Media, *J. Phys. Chem. A* **109**, pp. 1081–1093 (2005).

618. T. R. Chigwada and R. H. Simoyi, S-Oxygenation of Thiocarbamides II: Oxidation of Trimethylthiourea by Chlorite and Chlorine Dioxide, *J. Phys. Chem. A* **109**, pp. 1094–1104 (2005).

619. O. Olagunju, P. D. Siegel, R. Olojo and R. H. Simoyi, Oxyhalogen-Sulfur Chemistry: Kinetics and Mechanism of Oxidation of N-Acetylthiourea by Chlorite and Chlorine Dioxide, *J. Phys. Chem. A* **110**, pp. 2396–2410 (2006).

620. V. V. Udovichenko, P. E. Strizhak, A. Tóth, D. Horváth, S. Ning and J. Maselko, Temporal and Spatial Organization of Chemical and Hydrodynamic Processes. The System Pb^{2+}–Chlorite–Thiourea, *J. Phys. Chem. A* **112**, pp. 4584–4592 (2008).

621. G. Csekő, Y. Hu, Y. Song, T. R. Kégl, Q. Gao, S. V. Makarov and A. K. Horváth, Kinetic Evidence of Tautomerism of Thiourea Dioxide in Aqueous Acidic Solutions, *Eur. J. Inorg. Chem.* **11**, pp. 1875–1879 (2014).

622. Y. Hu, A. K. Horváth, S. Duan, G. Csekő, S. V. Makarov and Q. Gao, Mechanism Involving Hydrogen Sulfite Ions, Chlorite Ions, and Hypochlorous Acid as Key Intermediates of the Autocatalytic Chlorine Dioxide-Thiourea Dioxide Reaction, *Eur. J. Inorg. Chem.* **12**, pp. 5011–5020 (2015).

623. M. A. Salem, C. R. Chinake and R. H. Simoyi, Oxyhalogen-Sulfur Chemistry: Oxidation of Hydroxymethanesulfinic Acid by Chlorite, *J. Phys. Chem.* **100**, pp. 9377–9384 (1996).

624. G. Rábai and M. T. Beck, Oxidation of Thiourea by Iodate: A New Type of Oligo-oscillatory Reaction, *J. Chem. Soc. Dalton Trans.*, **14**, pp. 1669–1672 (1985).

625. J. Horváth, I. Szalai and P. De Kepper, An Experimental Design Method Leading to Chemical Turing Patterns, *Science* **324**, pp. 772–775 (2009).
626. A. F. Taylor and M. R. Tinsley, A Path to Patterns, *Nat. Chem* **1**, pp. 340–341 (2009).
627. J. Horváth, I. Szalai and P. De Kepper, Pattern Formation in the Thiourea-Iodate-Sulfite System: Spatial Bistability, Waves, and Stationary Patterns, *Physica D* **239**, pp. 776–784 (2010).
628. H. Liu, A. K. Horváth, Y. Zhao, X. Lv, L. Yang and Q. Gao, A Rate Law Model for the Explanation of Complex pH Oscillations in the Thiourea-Iodate-Sulfite Flow System, *Phys. Chem. Chem. Phys.* **14**, pp. 1502–1506 (2012).
629. G. Rábai, Z. Nagy and M. T. Beck, Quantitative Description of the Oscillatory Behavior of the Iodate-Sulfite-Thiourea System in CSTR, *React. Kin. Catal. Lett.* **33**, pp. 23–29 (1987).
630. J. F. Ojo, A. Otoikhian, R. Olojo and R. H. Simoyi, Oxyhalogen-Sulfur Chemistry: Nonlinear Oxidation Kinetics of Hydroxymethanesulfinic Acid by Acidic Iodate, *J. Phys. Chem. A* **108**, pp. 2457–2463 (2004).
631. Y. R. Du, S. Wang, J.-J. Lin, S.-M. Huang and H. Zhou, Nonlinear Oxidation of Thiourea in an Unbuffered Medium, *Chin. J. Chem.* **26**, pp. 1771–1779 (2008).
632. R. M. Noyes, A Generalized Mechanism for Bromate-Driven Oscillators Controlled by Bromide, *J. Am. Chem. Soc.* **102**, pp. 4644–4649 (1980).
633. B. P. Belousov, Periodically Acting Reaction and Its Mechanism, *Ref. Radiat. Med.*, pp. 145–147 (1959).
634. A. Pechenkin, B. P. Belousov and His Reaction, *J. Biosci.* **34**, pp. 365–371 (2009).
635. R. H. Simoyi, New Bromate Oscillator: The Bromate-Thiourea Reaction in a CSTR, *J. Phys. Chem.* **90**, pp. 2802–2804 (1986).
636. R. H. Simoyi, I. R. Epstein and K. Kustin, Kinetics and Mechanism of the Oxidation of Thiourea by Bromate in Acidic Solution, *J. Phys. Chem.* **98**, pp. 551–557 (1994).
637. S. B. Jonnalagadda, C. R. Chinake and R. H. Simoyi, Oxyhalogen-Sulfur Chemistry: Bromate Oxidation of 1-methyl-2-thiourea in Acidic Medium, *J. Phys. Chem.* **100**, pp. 13521–13530 (1996).
638. C. R. Chinake, R. H. Simoyi and S. B. Jonnalagadda, Oxyhalogen-Sulfur Chemistry: The Bromate-(Aminoimino)methanesulfinic Acid Reaction in Acidic Medium, *J. Phys. Chem.* **98**, pp. 545–550 (1994).
639. A. Otoikhian and R. H. Simoyi, Kinetics and Mechanism of Oxidation of N,N′- Dimethylaminoiminomethanesulfinic Acid by Acidic Bromate, *J. Phys. Chem. A* **112**, pp. 8569–8577 (2008).
640. S. B. Jonnalagadda, C. R. Chinake and R. H. Simoyi, Oxyhalogen-Sulfur Chemistry: Oxidation of Hydroxymethanesulfinic Acid by Bromate in an Acidic Medium, *J. Phys. Chem.* **99**, pp. 10231–10236 (1995).
641. M. H. El-Rafie, M. K. Zahran, K. F. El Tahlawy and A. Hebeish, A Comparative Study of the Polymerization of Acrylic Acid with Native and Hydrolyzed Maize Starches Using a Potassium Bromate–Thiourea Dioxide Redox Initiation System, *Polym. Degrad. Stab.* **47**, pp. 73–85 (1995).

642. E. S. Abdel-Halim, Preparation and Characterization of Poly(Acrylic Acid)-Hydroxyethyl Cellulose Graft Copolymer, *Carbohydr. Polym.* **90**, pp. 930–936 (2012).

643. T. L. Vigo, *Textile Processing and Properties. Preparation, Dyeing, Finishing and Performance*, 1st edn. Elsevier, Amsterdam (1994).

644. J. N. Chakraborty, *Fundamentals and Practices in Coloration of Textiles*, 1st edn. Woodhead Publ. India, New Delhi (2014).

645. U. Baumgarte, Über den Chemismus der Reduction von Köpenfarbstoffen, *Textileveredl.* **4**, pp. 821–832 (1969).

646. U. Baumgarte, Reduction and Oxidation Processes in Dyeing with Vat Dyes, *Melliand Textilber.* **68**, pp. 189–195 (1987).

647. U. Baumgarte, Entwiddungen bei Küpenfarbstoffen und ihrer Anwendung, *Melliand Textilber.* **56**, pp. 228–233 (1975).

648. U. Baumgarte, Developments in Vat Dyes and in Their Application, *Rev. Prog. Color.* **17**, pp. 29–38 (1987).

649. J. N. Etters, Efficient Use of Sodium Hydrosulfite in Commercial Vat Dyeing Processes, *Amer. Dyestuff. Rep.* **78**, pp. 18–26 (1989).

650. M. A. Kulandainathan, K. Patil, A. Muthukumaran and R. B. Chavan, Review of the Process Development Aspects of Electrochemical Dyeing: Its Impact and Commercial Application, *Color. Technol.* **123**, pp. 143–152 (2007).

651. M. Bozic and V. Kokol, Ecological Alternatives to the Reduction and Oxidation Processes in Dyeing with Vat and Sulphur Dyes, *Dyes Pigm.* **76**, pp. 299–309 (2008).

652. M. Sala and C. Gutierrez-Bouzán, Electrochemical Techniques in Textile Processes and Wastewater Treatment, *Int. J. Photoenergy*, **2012**, pp. 1–12 (2012).

653. T. Bechtold, E. Burtscher, A. Amann and O. Bobleter, Reduction of Dispersed Indigo Dye by Indirect Electrolysis, *Angew. Chem. Int. Ed.* **31**, pp. 1068–1069 (1992).

654. S. W. Dhawale, Thiosulfate. An Interesting Sulfur Oxoanion That is Useful in Both Medicine and Industry — But is Implicated in Corrosion, *J. Chem. Educ.* **70**, pp. 12–14 (1993).

655. A. Roessler and X. Jin, State of the Art Technologies and New Electrochemical Methods for the Reduction of Vat Dyes, *Dyes Pigm.* **59**, pp. 223–235 (2003).

656. T. Bechtold, E. Burtscher, G. Künnel and O. Bobleter, Electrochemical Reduction Processes in Indigo Dyeing, *J. Soc. Dyers Colour.* **113**, pp. 135–144 (1997).

657. S. K. Nicholson and P. John, The Mechanism of Bacterial Indigo Reduction, *Appl. Microbiol. Biotechnol.* **68**, pp. 117–123 (2005).

658. B. Semet, B. Säckingen and G. E. Grüningen, Eisen (II)-Salz-Komplexe als Alternative zu Hydrosulfit in der Küoen Färberei, *Melliand Textilber.* **76**, pp. 161–164 (1995).

659. C. R. Wasmuth, R. L. Donnell, C. E. Harding and G. E. Shankle, Kinetics of the Reduction of p-Phenylazobenzenesulphonic Acid by Sodium Dithionite in Alkaline Solution, *J. Soc. Dyers Colour.* **81**, pp. 403–405 (1965).

660. C. R. Wasmuth, C. Edwards and R. Hutcherson, Participation of the $SO_2 \bullet^-$ Radical Ion in the Reduction of p-Nitrophenol by Sodium Dithionite, *J. Phys. Chem.* **68**, pp. 423–425 (1964).

661. R. G. Rinker, T. P. Gordon, D. M. Mason, R. R. Sakaida and W. H. Corcoran, Kinetics and Mechanism of the Air Oxidation of the Dithionite Ion $(S_2O_4^{2-})$ in Aqueous Solution, *J. Phys. Chem.* **64**, pp. 573–581 (1960).

662. N. Srividya, G. Paramasivan, K. Seetharaman and P. Ramamurthy, Two-step Reduction of Indigo Carmine by Dithionite: A Stopped-flow Study, *J. Chem. Soc. Faraday Trans.* **90**, pp. 2525–2530 (1994).

663. M. Weiss, Thiourea Dioxide: A Safe Alternative to Hydrosulfite Reduction, *Amer. Dyestuff Rep.* **67**, pp. 35–38 (1978).

664. V. Olip, The Application of Formamidine Sulfinic Acid in the Textile Industry, *Melliand Textilber.* **80**, pp. 620–623 (1999).

665. T. K. Das, A. K. Mandavawalla, A. Lahiri and S. K. Datta, A Powerful and Safe Reducing Agent for Textile Applications, *Colourage* **31**, pp. 15–20 (1984).

666. J. Gacén, J. Cegarra and M. Caro, Wool Bleaching with Reducing Agent in the Presence of Sodium Lauryl Sulphate. Part 3 — Bleaching with Thiourea Dioxide, *J. Soc. Dyers Colour.* **107**, pp. 138–141 (1991).

667. H.-U. Suess, *Pulp Bleaching Today*, 1st edn. Walter de Gruyter, Berlin/N.Y. (2010).

668. P. Bajpai, *Pulp and Paper Industry: Chemicals*, 1st edn. Elsevier Science, Amsterdam (2015).

669. P. Bajpai, *Recycling and Deinking of Recovered Paper*, 1st edn. Elsevier Science, Amsterdam (2014).

670. P. W. Hart and A. W. Rudie, *The Bleaching of Pulp*, 5th edn. Tappi Press, Norcross (2012).

671. S. Robert, *The Chemistry of Lignin-Retaining Bleaching: Reductive Bleaching Agents, in Lignin and Lignans: Advances in Chemistry*, 1st edn. CRC Press, Boca Raton (2010).

672. T. Q. Hu, *Chemical Modification, Properties and Usage of Lignin*, 1st edn. Springer, Berlin (2002).

673. M. E. Ellis, *The Technology of Mechanical Pulp Bleaching, Chapter 2: Hydrosulfite (Dithionite) Bleaching, in: Pulp Bleaching — Principles and Practice*, 1st edn. Tappi Press, Atlanta (1996).

674. B. R. James, *New Chemistry, Including Pulp Bleaching Processes, of the Golden-Aged, Water-Soluble Compound, Tris(hydroxymethyl)phosphine, in Catalysis of Organic Reactions*, 1st edn. CRC Press, Boca Raton (2009).

675. C. W. Dence, *Section 3.2, Reactions of Sodium Dithionite with Chromophores in Lignin, in: Pulp Bleaching Principles and Practice*, 1st edn. Tappi Press, Atlanta (1996).

676. S. Hosoya, H. Hatakeyama and J. Nakano, Brightening of Lignin in Pulp with Sodium Hydrosulfite, *J. Jap. Wood Res. Soc.* **16**, pp. 140–144 (1970).

677. R. S. Pemberton, M. C. Depew, C. Heitner and J. K. S. Wan, Some Mechanistic Insights onto a Model Bleaching Process of Quinones by Bisulfite and Dithionite: An ESR-CIDEP Study, *J. Wood Chem. Technol.* **15**, pp. 65–83 (1995).

678. A. M. Devaney and R. G. Guess, Thiosulphate in Hydrosulfite Bleaching, *Pulp Paper Can.* **83**, pp. T60–T64 (1982).

679. T. Q. Hu, S. Johal, B. Yuen, T. Williams, D. A. Osmond and P. Watson, Thermomechanical Pulping and Bleaching of Blue-Stained Chips, *Pulp Paper Can.* **107**, pp. 38–45 (2006).

680. T. Q. Hu, T. Williams, S. Yazdi and P. Watson, Toward Overcoming the Brightness Ceiling of Mechanical Pulps Prepared from Blue-Stained Longepole Pine Chips, *Pulp Paper Can.* **110**, pp. 44–50 (2009).

681. D. M. Moiseev, B. R. James and T. Q. Hu, Characterization of Secondary and Primary (Hydroxymethyl)Phosphines and Their Oxidation Products: Synergism in Pulp-Bleaching, *Phosphorous Sulfur Silicon and Relat. Elem.* **187**, pp. 433–447 (2012).

682. B. R. James and F. Lorenzini, Developments in the Chemistry of Tris(hydroxymethyl)phosphine, *Coord. Chem. Rev.* **254**, pp. 420–430 (2010).

683. W. C. Mayer and C. P. Donofrio, Reductive Bleaching of Mechanical Pulp with Sodium Borohydride, *Pulp Paper Mag. Can.* pp. 157–166 (1958).

684. J. Blechschmidt, S. Wurdinger and G. Ziesenis, Single- and Multistep Bleaching of High-Yield Pulps with Hydrogen Peroxide and Formamidine Sulfinic Acid, *Papier* **45**, pp. 221–225 (1991).

685. C. Daneault, S. Robert and C. Leduc, Formamidine Sulfinic Acid as a Bleaching Chemical on Softwood TMP, *Res. Chem. Intermed.* **21**, pp. 521–533 (1995).

686. C. Daneault and C. Leduc, Bleaching of Mechanical Pulp with Formamidine Sulfinic Acid, *Cellul. Chem. Technol.* **28**, pp. 205–217 (1994).

687. L. Marchildon, C. Daneault, C. Leduc and M. M. Sain, Deinking Conditions for Yellow Directory Using Formamidine Sulfinic Acid as a Repulping Chemical, *Cellul. Chem. Technol.* **30**, pp. 473–482 (1996).

688. A. H. D. Vincent, C. Khong and E. Rizzon, FAS (Thiourea Dioxide) Bleaching of Recycled Pulp, *Appita J.* **50**, pp. 393–399 (1997).

689. J. D. Kronis, Optimum Conditions Play Important Role in Recycled Fiber Bleaching with FAS, *Pulp Paper* **9**, pp. 113–117 (1996).

690. S. Imamoglu, A. Karademir, E. Pesman, C. Aydemir and C. Atik, Effects of Flotation Deinking on the Removal of Main Colors of Oil-based Inks from Uncoated and Coated Office Papers, *Bio. Resources* **8**, pp. 45–58 (2013).

691. A. Castellan, J. H. Zhu, N. Colombo, A. Nourmamode, R. S. Dawson and L. Dunn, An Approach to Undestanding the Mechanism of Protection of Bleached High-Yield Pulps Against Photoyellowing by Reducing Agents Using the Lignin Model Dimer: 3,4-Dimetoxy-α-(2-methoxyphenoxy)-acetophenone, *J. Photochem. Photobiol. A: Chem.* **58**, pp. 263–273 (1991).

692. I. M. Kolthoff and E. J. Meehan, Redox Recipes. X. A Recipe with Ferric Versenate, Sodium Dithionite, and Hydroperoxide as Activating System (Veroxathionite Recipe), *J. Appl. Polym. Sci.* **1**, pp. 200–211 (1959).

693. J. M. Mitchell, R. Spolsky and H. L. Williams, Use of Alternative Activators in 41 °F. Polymerization Recipes, *Ind. Eng. Chem.* **41**, pp. 1592–1599 (1949).

694. A. K. Prince and R. D. Spitz, Synthetic Rubber Production. Chelating Agents in Sulfoxylate Polymerization,, *Ind. Eng. Chem.* **52**, pp. 235–238 (1960).

695. R. W. Brown, C. V. Bawn, E. B. Hansen and L. H. Howland, Sodium Formaldehyde Sulfoxylate in GR-S Polymerization, *Ind. Eng. Chem.* **46**, pp. 1073–1080 (1954).

696. R. Kerber and A. Effmert, Zur Kinetik des Redoxsystems Rongalit/Wasserstoffperoxyd I. *Makromol. Chem.* **50**, pp. 220–231 (1961).

697. R. Kerber and X. Gregory, Zur Kinetik des Redoxsystems Rongalit/Wasserstoffperoxyd II. *Makromol. Chem.* **68**, pp. 100–119 (1963).

698. V. C. C. Lin, F. Patat and E. Staude, Die Redoxreaktion Wasserstoffperoxid/Rongalit und ihre katalytische Wirkung auf die Polymerisation von Acrylnitrit, *Angew. Makromol. Chem.* **8**, pp. 28–40 (1969).

699. H. Minato and M. Iwakawa, Generation of Free Radicals at Subzero Temperatures. IV. On the Mechanism of Generation of Free Radicals from the Sodium Formaldehyde Sulfoxylate–Iron–Hydroperoxide System, *Polym. J.* **6**, pp. 234–237 (1974).

700. E. S. Daniels, V. L. D. Dimonie, M. S. El-Asser and J. W. Vaederhoff, Preparation of ABS (Acrylonitrile/Butadiene/Styrene) Latexes Using Hydroperoxide Redox Initiators, *J. Appl. Polym. Sci.* **41**, pp. 2463–2477 (1990).

701. G. Kojima and M. Hisasue, Die Emulsionscopolymerisation von Tetrafluorethylen mit Propylen bei niedrigen Temperaturen,, *Makromol. Chem.* **182**, pp. 1429–1439 (1981).

702. C. C. Wang, J. F. Kuo and C. Y. Chen, Dimerizations of Acrylate Monomers with Sodium Hydroxymethanesulfinate and Characterization of the Products, *Macromol. Chem. Phys.* **195**, pp. 1493–1502 (1994).

703. C. C. Wang, N. S. Yu, C. Y. Chen and J. F. Kuo, Kinetic Study of the Miniemulsion Polymerization of Styrene, *Polymer* **37**, pp. 2509–2516 (1996).

704. W. Arayapranee, P. Prasassarakich and G. L. Rempel, Synthesis of Graft Copolymers from Natural Rubber Using Cumene Hydroperoxide Redox Initiator, *J. Appl. Polym. Sci.* **83**, pp. 2993–3001 (2002).

705. N. Kohut-Svelko, R. Pirri, J. M. Asua and J. R. Leiza, Redox Initiator Systems for Emulsion Polymerization of Acrylates, *J. Polym. Sci.* **47**, pp. 2917–2927 (2009).

706. A. L. Elorza and J. M. A. González, *Aqueous Hybrid Polyurethane-Acrylic Adhesives*, U.S. Patent 2012/0208014 A1. Universidad del Pais Vasco, Leioa-Vizcaya, Espana (2012).

707. L. Martin, G. Gody and S. Perrier, Preparation of Complex Multiblock Copolymers via Aqueous RAFT Polymerization at Room Temperature, *Polym. Chem.* **6**, pp. 4875–4886 (2015).

708. Z. Liu, Y. Han, C. Zhou, M. Zhang, W. Li, H. Zhang, F. Liu and W.-J. Liu, Seeded Emulsion Polymerization of Butyl Acrylate Using a Redox Initiator System: Kinetics and Mechanism, *Ind. Eng. Chem. Res.* **49**, pp. 7152–7158 (2010).

709. C. Ge, Y. Wu and J. Xu, Stability and Optimum Polymerized Condition of Polysiloxane-Polyacrylate Core-Shell Polymer, *Adv. Polym. Technol.* **29**, pp. 161–172 (2010).

710. M. Radenkov, A. Topliyska, P. Kyulanov and P. Radenkov, Activity of a Modified Redox System for Polymerization of Unsaturated Polyester Resins, *Polym. Bull.* **52**, pp. 275–282 (2004).

711. V. Percec, A. V. Popov, E. Ramirez-Castillo, J. F. J. Coelho and L. A. Hinojosa-Falcon, Non-transition Metal-Catalyzed Living Radical Polymerization of Vinyl Chloride Initiated with Iodoform in Water at 25 °C, *J. Polym. Sci. A.: Polym. Chem.* **42**, pp. 6267–6282 (2004).

712. V. Percec, A. V. Popov, E. Ramirez-Castillo and O. Weichold, Acceleration of the Single Electron Transfer-Degenerative Chain Transfer Mediated Living Radical Polymerization (SET-DTLRP) of Vinyl Chloride in Water at 25 °C, *J. Polym. Sci. A.: Polym. Chem.* **42**, pp. 6364–6374 (2004).

713. V. Percec, A. V. Popov, E. Ramirez-Castillo and J. F. J. Coelho, Single-Electron-Transfer/Degenerative-Chain-Transfer Mediated Living Radical Polymerization (SET-DTLRP) of Vinyl Chloride Initiated with Methylene Iodide and Catalyzed by Sodium Dithionite, *J. Polym. Sci. A: Polym. Chem.* **43**, pp. 773–778 (2005).

714. V. Percec, A. V. Popov, E. Ramirez-Castillo, J. F. J. Coelho and L. A. Hinojosa-Falcon, Phase Transfer Catalyzed Single Electron Transfer-Degenerative Chain Transfer Mediated Living Radical Polymerization (PTC-SET-DTLRP) of Vinyl Chloride Catalyzed by Sodium Dithionite and Initiated with Iodoform in Water at 43 °C, *J. Polym. Sci. A: Polym. Chem.* **43**, pp. 779–788 (2005).

715. V. Percec, E. Ramirez-Castillo, A. V. Popov, L. A. Hinojosa-Falcon and T. Guloashvili, Ultrafast Single-Electron-Transfer/Degenerative-Chain-Transfer Mediated Living Radical Polymerization of Acrylates Initiated with Iodoform in Water at Room Temperature and Catalyzed by Sodium Dithionite, *J. Polym. Sci. A: Polym. Chem.* **43**, pp. 2178–2184 (2005).

716. V. Percec, V. A. Popov, E. Ramirez-Castillo and L. A. Hinojosa-Falcon, Synthesis of Poly(Vinyl Chloride)-*b*-Poly (2-Ethylhexyl acrylate)-*b*-poly (Vinyl Chloride) by the Competitive Single-Electron-Transfer/Degenerative-Chain-Transfer Mediated Living Radical Polymerization of Vinyl Chloride Initiated from α, ω-di (iodo) poly (2-Ethylhexyl Acrylate) and Catalyzed with Sodium Dithionite in Water, *J. Polym. Sci. A: Polym. Chem.* **43**, pp. 2276–2280 (2005).

717. J. F. J. Coelho, A. M. F. P. Silva, A. V. Popov, V. Percec, M. V. Abreu, P. M. Goncalves and M. H. Gil, Single Electron Transfer/Degenerative-Chain-Transfer Mediated Living Radical Polymerization of N-butyl Acrylate Catalyzed by $Na_2S_2O_4$ in Water Media,, *J. Polym. Sci. A: Polym. Chem.* **44**, pp. 2809–2825 (2006).

718. J. F. J. Coelho, E. Y. Carvalho, D. S. Marques, A. V. Popov, P. M. Goncalves and M. H. Gil, Synthesis of Poly(lauryl acrylate) by Single-Electron

Transfer/Degenerative Chain Transfer Living Radical Polymerization Catalyzed by $Na_2S_2O_4$ in Water, *Macromol. Chem.* **208**, pp. 1218–1227 (2007).

719. J. F. J. Coelho, J. Góis, A. C. Fonseca, R. A. Carvalho, A. V. Popov, V. Percec, P. M. Goncalves and M. H. Gil, Synthesis of Poly(ethyl acrylate) by Single Electron Transfer-Degenerative Chain Transfer Living Radical Polymerization in Water Catalyzed by $Na_2S_2O_4$, *J. Polym. Sci. A: Polym. Chem.* **46**, pp. 421–432 (2008).

720. J. F. J. Coelho, J. Góis, A. C. Fonseca, R. A. Carvalho, A. V. Popov, V. Percec and M. H. Gil, Synthesis of Poly (2-methoxyethyl acrylate) by Single Electron Transfer-Degenerative Chain Transfer Living Radical Polymerization Catalyzed by $Na_2S_2O_4$ in Water, *J. Polym. Sci. A: Polym. Chem.* **47**, pp. 4454–4463 (2009).

721. J. F. J. Coelho, E. Y. Carvalho, D. S. Marques, A. V. Popov, V. Percec and M. H. Gil, Influence of the Isomeric Structures of Butyl Acrylate on Its Single-Electron Transfer-Degenerative Chain Transfer Living Radical Polymerization in Water Catalyzed by $Na_2S_2O_4$, *J. Polym. Sci. A: Polym. Chem.* **46**, pp. 6452–6551 (2008).

722. J. R. Góis, N. Rocha, A. V. Popov, T. Guliashvili, K. Matyjaszewski, A. C. Serra and J. F. J. Coelho, Synthesis of Well-defined Functionalized Poly(2-(Diisopropylamino)ethyl Methacrylate) Using ATRP with Sodium Dithionite as a SARA Agent, *Polym. Chem.* **5**, pp. 3919–3928 (2014a).

723. J. R. Góis, D. Konkolewic, A. V. Popov, T. Guliashvili, K. Matyjaszewski, A. C. Serra and J. F. J. Coelho, Improvement of the Control over SARA ATRP of 2-(diisopropylamino)ethyl methacrylate by Slow and Continuous Addition of Sodium Dithionite, *Polym. Chem.* **5**, pp. 4617–4626 (2014b).

724. C. M. R. Abreu, A. C. Serra, A. V. Popov, K. Matyjaszewski, T. Guliashvili and J. F. J. Coelho, Ambient Temperature Rapid SARA ATRP of Acrylates and Methacrylates in Alcohol/Water Solutions Mediated by Mixed Sulfites/Cu(II)Br$_2$ Catalytic System, *Polym. Chem.* **4**, pp. 5629–5636 (2013).

725. G. Gelbard, O. Louis-André and O. Cherkaoui, Reductions with Polymer Supported Dithionite Anions: Regioselectivity in Conjugated Systems, *React. Polym.* **15**, pp. 111–119 (1991).

726. A. Hebeish, M. H. El-Rafie, A. Waly and A. Z. Moursi, Graft Copolymerization of Vinyl Monomers onto Modified Cotton. IX. Hydrogen Peroxide-Thiourea Dioxide Redox System Induced Grafting of 2-Methyl-5-vinylpyridine onto Oxidized Celluloses, *J. Appl. Polym. Sci.* **22**, pp. 1853–1866 (1978).

727. A. Waly, N. Y. Abou-Zeid, E. A. El-Alfy and A. Hebeish, Polymerization of Glycidyl Methacrylate, Methacrylic Acid, Acrylamide and their Mixtures with Cotton Fabric Using Fe^{2+}–Thiourea Dioxide–H_2O_2 Redox System, *Angew. Makromol. Chem.* **103**, pp. 61–76 (1982).

728. N. Y. Abou-Zeid, A. Waly, E. A. El-Alfy and A. Hebeish, Fe^{2+}–Thiourea Dioxide–H_2O_2–Induced Polymerization of Glycidyl Methacrylate and Its Mixtures with Acrylamide, Acrylonitrile, Butylmethacrylate, or Styrene with Cotton Fabric, *J. Appl. Polym. Sci.* **27**, pp. 2105–2117 (1982).

729. N. Y. Abou-Zeid, A. Waly, A. Higazy and A. Hebeish, Fe^{2+}–Thiourea Dioxide-H_2O_2–Induced Polymerization of Various Vinyl Monomers with Flax Fibres, *Angew. Makromol. Chem.* **143**, pp. 85–100 (1986).

730. S. A. Abdel-Hafiz, Potassium Permanganate/Thiourea Dioxide Redox System-Induced Grafting of Methacrylic Acid onto Loomstate Cotton Fabric, *J. Appl. Polym. Sci.* **58**, pp. 2005–2011 (1995).

731. E. S. Abdel-Halim and S. S. Al-Deyab, Preparation of Poly(acrylic acid)/Starch Hydrogel and Its Application for Cadmium Ion Removal from Aqueous Solutions, *React. Funct. Polym.* **75**, pp. 1–8 (2014).

732. V. Percec, A. V. Popov and E. Ramirez-Castillo, Single-Electron-Transfer/Degenerative-Chain-Transfer Mediated Living Radical Polymerization of Vinyl Chloride Catalyzed by Thiourea Dioxide/Octyl Viologen in Water/Tetrahydrofuran at 25°C, *J. Polym. Sci. A: Polym. Chem.* **43**, pp. 287–295 (2005).

733. N. N. Vlasova, Y. N. Pozhidaev, O. Y. Raspopina, L. I. Belousova and M. G. Voronkov, Polyorganylsilsesquioxanes Containing Carbofunctional Groups $(NH)_2C(SO_2)$. Synthesis and Sorption Properties, *Russ. J. Gen. Chem.* **69**, pp. 1391–1394 (1999).

734. I. V. Karandi and M. M. Buzlanova, Determination of Organic Disulfides by Potentiometry Using Formamidinesulfinic Acid as the Reducing Agent, *J. Anal. Chem.* **54**, pp. 753–755 (1999).

735. N. K. Pshenitsin, I. V. Prokofeva and A. V. Bukanova, Formamidinesulfinic Acid in Analytical Chemistry of Platinum Metals. 1. Determination of Rhodium, *Zh. Anal. Khim.* **18**, pp. 761–764 (1963).

736. I. V. Prokofeva and A. V. Bukanova, Formamidinesulfinic Acid in Analytical Chemistry of Platinum Metals. 1. Separation of Rhodium and Iridium, *Zh. Anal. Khim.* **20**, pp. 598–609 (1965).

737. T. Rudman, *The Photographers Toning Book. The Definitive Guide*, 1st edn. Amphoto, New York (2003).

738. A. Mallick, D. Lai and S. Roy, Autonomous Movement Induced in Chemically Powered Active Soft-Oxometalates Using Dithionite as Fuel, *New J. Chem.* **40**, pp. 1057–1062 (2016).

739. W. Gao, W. Qi, J. Lai, L. Qi, S. Majeed and G. Xu, Thiourea Dioxide as Unique Eco-friendly Coreactant of Luminol Chemiluminescence for Sensitive Detection of Luminol, Thiourea Dioxide and Cobalt Ions, *Chem. Commun.* **51**, pp. 1620–1623 (2015).

740. http://www.swedishclub.com/upload/ Loss_Prev_Docs/Cargo/Misdeclared_or_undeclared_DGC_Letter1-07-2.pdf.

741. http://www.figuk.org.uk/conferences/2014/chris_foster.pdf.

Index

A

Acetylthiourea
    reaction with chlorite, 132
Acrylates
    polymerization, 148–150,
      152–154
Alcohols
    synthesis, 75–77, 81–82, 84
    oxidation, 104
Aldehydes
    reactions with sodium dithionite,
      80–82,
    reactions with sodium
      hydroxymethanesulfinate, 83
    reactions with thiourea dioxide,
      84–85
    catalytic reaction with
      dimedone, 102
    α,β-unsaturated monoterpenic,
      reduction, 82
Allyl alcohol
    synthesis, 75–77
    mechanism of formation, 76–77
Allyl thiourea
    reaction with singlet oxygen, 18
Amines
    synthesis from aromatic nitro
      compounds, 75, 87, 89, 91
    reactions with
      N-phenylthiourea trioxide,
      61
      thiourea dioxide, 62
      thiourea trioxide, 62
Aminoiminomethanesulfenic acid, see
    thiourea monoxide
Aminoiminomethanesulfinic acid, see
    thiourea dioxide

Aminoiminomethanesulfonic acid, see
    thiourea trioxide
α-Aminomethanesulfinates
    synthesis, 10–11
Aminotetrazoles
    synthesis, 90
Ammonia,
    reaction with thiourea dioxide,
      62
Antraquinone derivatives
    reduction, 139–141
Arginine, 59–60
Azo compounds
    determination, 92

B

Benzils
    reduction, 83–84
Benzimidazoles
    synthesis, 92
Benzosulfones
    synthesis, 69–70
Benzotriazoles
    synthesis, 88–89
Benzotriazoles 1-oxides
    synthesis, 88
Biginelli condensation, 97–98
    plausible mechanism, 99
Bitumen
    modification with thiourea
      dioxide, 157
Bleached chemithermomechanical
   pulp
    discoloration, 146
Bromate
    reactions with
      N,N'-dimethylthiourea
      dioxide, 137

N,N'-dimethylthiourea
trioxide, 137
sodium hydroxymethanesul-
finate,
137
N-methylthiourea, 136
thiourea, 129, 132,136
trimethylthiourea, 136
Bruggolite FF6 11
usage in polymerization of
acrylates, 149
Bruggolite FF7
usage in polymerization of
acrylates, 149

C

Cannizzaro reaction, 82
Carboxyl graphene
synthesis, 113
Chemiluminescence, 158
Chitosan, guanidinylated
synthesis, 65–66
Chlorine dioxide
reactions with
thiourea, 130
thiourea dioxide, 133
Chlorite
reactions with
acetylthiourea, 132
sodium hydroxymethanesul-
finate, 133
N-phenylthiourea, 132
tetramethylthiourea, 132
thiourea, 129–132
thiourea dioxide, 132
trimethylthiourea, 132
Cobalamins
reduction, 122
Cobalt
cage complexes with nitro groups
reduction, 127
phthalocyanine
catalyst, 104–105
tetrasulfophthalocyanine
catalyst of reduction of
nitrite and nitrate,
119–122
reduction, 119–120
Combined metal and organocatalysis,
104–105

Coumarins
synthesis, 100
Cumene hydroperoxide
usage in polymerization
processes, 150
Cyanamide, 12, 18, 44

D

Decamethylcyclopentasiloxane, 143
Decrolin (see zinc
hydroxymethanesulfinate)
Diaryl diselenides
synthesis, 80
Diaryl ditellurides
synthesis, 80
Dibromoalkanes, 75
Di-n-butyl telluride
synthesis, 74
Dicyandiamide, 12,13
Diethylaminomethanesulfinate,
sodium
synthesis, 10
N,N'-Diethylthiourea
reaction with hydrogen peroxide,
14
Dihydropyrido[2,3-d]pyrimidine-2,4-
diones
synthesis, 98–99
3,4-Dihydropyrimidinones
synthesis, 97–99
Diisopropylbenzene hydroperoxide
usage in polymerization
processes, 147
Dimedone
catalytic reaction with
aldehydes, 102
Dimethylaminomethanesulfinate,
sodium
synthesis, 10
N,N'-Dimethylthiourea
reaction with hydrogen peroxide,
14
dioxide
decomposition in alkaline
solutions, 44
reaction with
bromate, 135
iodine, 135
selected bond distances, 29
trioxide, hydrate

selected bond distances, 29
Dinitrosyl iron complexes
    reaction with sodium dithionite,
        127
Disulfides,
    determination, 157
Dithionate, sodium
    structure, 22
Dithionite complexes
    itterbium
        structure, 25
    rhodium
        photochromic
            transformation, 23–25
        structure, 23–25
    samarium
        structure, 26
    uranium
        structure, 25–26
Dithionite dianion
    aromaticity, 26–27
Dithionites
    lithium
        purity, 33
        structure, 23
        synthesis, 7
    potassium
        purity, 33
    sodium
        anaerobic decomposition in
            acidic aqueous solutions,
            39–42
        anaerobic decomposition in
            alkaline aqueous
            solutions, 42–43
        determination, 33–37
        density of aqueous
            solutions, 33
        oscillatory decomposition in
            a continuous-flow stirred
            tank reactor, 41
        purity, 33
        reactions with
            aldehydes, 80–82
            alkyl dihalides, 66
            alkyl halides, 66
            allyl halides, 69
            anthraquinone
                derivatives, 139–142
            benzyl halides, 69, 71

α-chloromethy
    lnaphtalene, 66
cobalamins, 122
cobalt complexes,
    including tetrasul-
    fophthalocyanine,
    116–119, 121–122,
    128
graphene (graphite)
    oxides, 110–111
hydrogen peroxide,
    46–47
indigocarmine,
    141–142
iron complexes,
    including tetrasul-
    fophthalocyanines,
    117–120, 127
ketones, 81–83
lignin, 144
manganese complexes,
    115–116
Ni^{2+} 124
nicotinamide adenine
    dinucleotide, 94–95
nitrate, 119–120
nitrite, 119
nitrogen functions in
    organic compounds,
    92–93
nitrocompounds, 92
α-nitrosulfones, 73
organofluorine
    compounds, 54–57
oxygen, 44–47
proteins and peptides,
    116, 118–119,
    126–127
pyridinium salts, 96
selenious acid, 80
soft-oxometalates, 158
sulfur dyes, 139
vat dyes, 139–142
vinyl compounds, 66
vinylsulfones, 71–72
viologens, 93–94
solubility, 33
structure, 21–23
synthesis, 5–7
thermal decomposition, 33

toxicity, 33
usage
    as a bleaching agent in
        paper industry,
        143–145
    for dechlorination of
        organic compounds,
        120
    for iron stain removal,
        126
    for preparation of Fe/FeS
        nanoparticles, 123–124
    for preparation of nano
        zero valent iron,
        122–123
    for preparation of nickel
        sulfide, 124
    in polymerization of
        acrylates, 153
    in polymerization of
        vinyl chloride, 150
    on exchange resins, 153
viscosity of aqueous solutions, 33
tetraethylammonium
    synthesis, 7
tin
    structure, 21
    synthesis, 7
Dushman reaction, 134

E

Ethylenethiourea
    (imidazolidine-2-thione)
    reaction with hydrogen peroxide,
        19
    monoxide
        synthesis, 13

F

Ferrate
    reaction with thiourea, 18
Fluorinated extended porphyrins
    synthesis, 55–56
Fluorous seleninic acid
    synthesis, 79
Formamidine, 12
Formamidinedisulfide
    synthesis, 12
    oxidation, 13, 17
    stability, 17

Formamidinesulfenic acid, see
    thiourea monoxide
Formamidinesulfinic acid, see
    thiourea dioxide
Formamidinesulfonic acid, see
    thiourea trioxide

G

Graphene, 107–109
    synthesis from graphite, 108–109
    chemically reduced, 111–115
Graphene oxide
reduction, 107, 109–115
Graphene hydrogels
    preparation, 113–114
Graphite, 107–109
Graphite oxide
    model, 108
    reduction, 107
Guanidines
    synthesis, 59–66
    triaryl
        synthesis, 18
Guanidinoacetic acid
    synthesis, 60

H

Hydrogen peroxide
    determination, 92
    reaction with
        sulfoxylate, 48
        sulfur dioxide anion radical,
            48
        thioureas, 11–17, 142–143
Hydroxyl radical
    reaction with thiourea and
        tetramethylthiourea, 18
α-Hydroxyalkanesulfinates
    synthesis, 9–11, 80–81
    usage in textile industry,
        139–140
α-Hydroxyarenesulfinates
    synthesis, 80–81
α-Hydroxyethanesulfinate
    synthesis, 10
    reactions with anthraquinone
        derivatives, 140–141
Hydroxymethanesulfinate,
    calcium (Rongalite H)
        synthesis, 10

sodium (Rongalite)
  aerobic decomposition in
    acidic solutions, 47–48
  anaerobic decomposition in
    acidic solutions, 43
  1,4- and 1,2-benzoquinones,
    66–67
  determination, 37
  reactions with
    acrylates, 148–149
    alkenyl bromides, 70
    aromatic aldehydes, 83
    benzils, 83
    benzyl bromides, 69
    benzyl chloride, 66
    bromate, 137
    chlorite, 133
    iodate, 135
    iodine, 37, 134–135
    cobalamins, 122
    cobalt tetrasulfophthalo-
      cyanine,
      121–122
    diaryl disulfides and
      diselenides, 78, 80
    α,α'-dibromo-o-xylene,
      69
    4-(dimethylamino)-1,3-
      diphenylbutan-2-one
      hydrochloride, 67
    ferric ethylenediaminete-
      traacetate,
      147–148
    α-haloketones, 83
    heterocyclic-fused
      dibromides, 74
    iron tetrasulfophthalo-
      cyanine and its
      μ-nitrido- and
      μ-oxodimers, 121–122
    naphtoquinone, 66–67
    nitrite, 121–122
    nitrophenylazo dyes, 89
    organofluorine
      compounds, 54, 57
    selenium, 76
    tellurium, 74, 76
    tellurium and n-butyl
      bromide, 74
    vat dyes, 141
  vinyl compounds, 66
  solubility, 37
  structure, 27
  synthesis, 9–10
  thermal decomposition, 37
  toxicity, 37
  usage
    as a discharging agent,
      141–142
    for preparation of metal
      sulfides, 124
    in chemical metallization
      of polyacrylonitrile,
      125
    in polymerization of
      acrylonitrile-
        butadiene-styrene
        latexes, 148
      styrene, 149
      unsaturated polyester
        resins, 150
zinc (Decrolin)
  reactions with organofluorine
    compounds, 54
  structure, 28
  synthesis, 10
Hydroxymethanesulfonate, sodium
  reaction with iodine, 135

I

Imidazolidine-2-thione, see
    ethylenethiourea
Imines
  hydrolysis, 104–105
Indigocarmine
  reaction with sodium dithionite,
    141
Iodate
  reactions with
    sodium hydroxymethanesul-
      finate,
      135
    thiourea, 129, 132–134
    thiourea dioxide, 134
Iodine
  reactions with
    N,N'-dimethylthiourea
      dioxide, 134–135
    hydroxymethanesulfinate,
      37, 134–135

hydroxymethanesulfonate,
135
thiourea dioxide, 38,
134–135
thiourea trioxide, 38,
134–135
Ionic liquids
in synthesis of dibenzylsulfones,
71
Iridium,
determination, 157
Iron
dinitrosyl complexes
reduction, 127
iron/iron sulfides nanoparticles
using for reduction of
chlorinated organic
compounds, 123–124
nano zero valent
synthesis, 123
usage for groundwater
remediation, 122
tetrasulfophthalocyanine
reduction, 121
catalyst of reduction of
nitrite, 121

J

Julia reaction, 71

K

Ketones
reactions with sodium dithionite,
80, 82–83, 86
reactions with thiourea dioxide,
84–87
α, β-unsaturated monoterpenic,
reduction, 82

L

Langlois' reagent (see sodium
trifluoromethanesulfinate)

M

Marine sponge diterpenoid
nakamurol A
synthesis, 77
Melamine

product of thiourea trioxide
decomposition, 44
Metabisulfite, sodium
structure, 22
Metal nanoparticles
synthesis, 125
Metallization of polyacrylonitrile, 125
Metalloporphyrins
reduction, 115–117
Methemerythrin
reduction, 118
Metmyoglobin
reduction, 116, 118, 121
N-Methylthiourea
reaction with bromate, 136
N-Methylthiourea dioxide, hydrate
decomposition in alkaline
solutions, 44
selected bond distances, 29
2-Methyl-5-vinylpyridine
grafting onto oxidized celluloses,
153–154
Methyl viologen
reduction, 82, 92
as an electron transfer catalyst,
82
Michael addition, 78

N

Nicotinamide adenine dinucleotide
reduction, 92
Nitrate
reduction, 119–120
Nitric acid, esters
determination, 92
Nitrite
reduction, 119–121
Nitro compounds
determination, 91–92
reduction, 92
synthesis, 73
Nitrophenols
determination, 92
Nitroso compounds
determination, 92
Nitrostilbenes
determination, 92

O

Olefinic sulfones
    synthesis, 70
    mechanism of formation, 71
Olefins
    synthesis from sulfones, 71
    synthesis from dibromoalkanes,
      75
Organocatalysis, 96–105
Organofluorine compounds, 53–59

P

Peracetic acid
    reaction with thiourea, 12
Perfluoroalkanesulfinates
    synthesis, 53
Periodate
    reaction with iodate, 136
Peroxomonosulfate
    reaction with thiourea, 16
Phase transfer catalysis, 69–70,
    73–74, 76, 82, 84–85, 93, 152
N-Phenylthiourea
    reaction with chlorite, 129, 132
    monoxide
        synthesis, 13
    trioxide
        in guanidine synthesis, 60,
          61
Polyacrylonitrile
    metallization, 125
Polyfluoroalkanesulfinates
    synthesis, 53
Polyfluoroalkanesulfonates
    synthesis, 53
Polyfluoroalkyl radicals, 54–55
Polyhydroquinolines
    synthesis, 102
Polymerization, 147–155
Polysulfones,
    synthesis, 66
Polytellurides
    synthesis, 75–76
Polythionite, 8, 51
Proteins
    reduction, 116, 118–119
Pyrimidines
    synthesis, 90
Pyridinium salts
    reduction, 96

Pyrroles
    synthesis, 90

R

Reductive dechlorination, 120
Rhodium
    determination, 157
Rongal A, 10
Rongalite (see sodium
    hydroxymethanesulfinate)

S

Selena-fatty acids
    synthesis, 79
Selenides
    alkyl
        synthesis, 80
    aryl
      synthesis, 80
    β-amino/β-hydroxy
      synthesis, 78
    Z)-1-alkenyl
      synthesis, 79
      mechanism of formation, 79
Selenium
    synthesis, 80
Selenoesters
    synthesis, 79
Single-electron-transfer mechanism,
    73, 150–152
Singlet oxygen
    reactions with thiourea and allyl
      thiourea, 18
Smiles rearrangement, 89
Sodium telluride
    synthesis, 74
    reactions with
      alkyl halides, 80
      aromatic nitro compounds,
        75
      bis(trimethyl-silyl)-1,3-
        butadiyne, 74
      β,β'-dichlorodiethyl sulfide,
        74
      chloromethyloxiranes, 75

Sodium trifluoromethanesulfinate
(Langlois' reagent)
    synthesis, 56
    usage for trifluoromethylation,
        56–57
Soft-oxometalates, 158
Sulfides organic
    oxidation, 102–103
    synthesis, 73–74
    β-amino/β-hydroxy
        synthesis, 78
    β-hydroxy
        synthesis, 77–78
        mechanism of formation, 78
    β-sulfido carbonyl compounds
        synthesis, 77–78
    (Z)-1-alkenyl
        synthesis, 79
        mechanism of formation, 79
Sulfides inorganic
    copper
        synthesis, 124
    nickel
        synthesis, 124
    silver
        synthesis, 124
    zinc
        synthesis, 124–125
Sulfinatodehalogenation, 53–55
Sulfones,
    synthesis, 66–69, 71
    dibenzyl
        synthesis, 66–69, 71
Sulfoxides
    synthesis, 102–103
Sulfoxylate
    formation in the course of
        thiourea dioxide
        decomposition in alkaline
        solutions, 48
    reactions with
        anthraquinone derivatives,
            141
        iron
            tetrasulfophthalocyanine
            and its μ-nitrido- and
            μ-oxodimers, 121–122
        perfluoroalkyl iodides, 58
        hydrogen peroxide, 48

        oxygen, 48
        superoxide, 48
Sulfur, active, 39
Sulfur dioxide
    biological role, 48–49
    reduction in non-aqueous
        solutions, 51–52
Sulfur dioxide anion radical
    determination, 49–50
    equilibrium with dithionite
        dianion, 21
    product of sulfur dioxide
        reduction, 48–49
    reactions with
        cobalamin (I) 122
        metal complexes, 116–118
        methemerythrin, 49
        nitro compounds, 49
        organofluorine compounds,
            54, 56
        oxygen, 45, 48
        hydrogen peroxide, 48
        superoxide, 48
    structure, 50
Sulfur dyes
    reduction, 139
Sultines
    synthesis, 74
Superoxide
    reactions with
        sulfoxylate, 48
        sulfur dioxide anion radical,
            48
        thiourea and
            diarylthioureas, 18

T

Tellurophene
    synthesis, 74
Tert-butyl hydroperoxide
    reactions with
        alcohols, 104
        sulfides, 102–104
        trifluoromethanesulfinate,
            56
    usage in polymerization
        processes, 149–150
Tetraalkylthiourea trioxides
    synthesis, 14

Tetrahydroselenophene
  synthesis, 76
Tetramethylformamidinedisulfide
  synthesis, 14
Tetramethylthiourea
  reactions with
    chlorite, 13, 129, 132
    bromate and bromine, 13,
      136
    monoxide
      formation, 13, 133, 136–137
    trioxide
      synthesis, 14
Tetramethylurea, 13
1,4-Thiatellurane
  synthesis, 74
Thermomechanical pulp
  bleaching, 145
Thioesters
  synthesis, 79
Thiols
  synthesis, 73–74
Thiourea
  C-S bond length, 28
  organocatalyst, 97
  oxidation, 11–20
  reactions with
    bromate, 129
    chlorine, 129
    chlorite, 129
    hydrogen peroxide, 11–17,
      142–143
    hydroxyl radical, 18
    peracetic acid, 12
    periodate, 136
    singlet oxygen, 18
  starting material in guanidine
    synthesis, 60
  diaryl
    reaction with superoxide, 18
Thiourea dioxide
  carbenoid structure, 29–30
  catalyst of synthesis of
    coumarin, 100
    3,4-dihydropyrimidinones,
      97–98
    dihydropyrido[2,3-
      d]pyrimidine-2,4-diones,
      98–99
    heterocycles, 98, 100, 101

polyhydroquinolines, 102
pyran derivatives, 100–108
sulfoxides, 102–103
catalyst of
  oxidation of alcohols, 104
  hydrolysis of imines,
    104–105
clusters in water, 31
complex with polyethyleneglycol,
  97–98
crystal structure at ambient and
  high pressure, 31–32
decomposition in aqueous
  solutions, 43, 158
determination, 38
DFT calculations, 30, 31
hydrogen bonds, 30–32, 97
oxygen release, 51–52
reactions with
  aldehydes, 85
  amines, 62
  amino acids, 60
  aromatic nitrocarbonyls, 85
  arylselenium trihalides, 80
  aryltellurium trihalides, 80
  azo compounds, 87
  azoxy compounds, 87
  bromate, 137
  Cd^{2+} 125
  chlorine dioxide, 133
  chlorite, 132
  cobalamins, 122
  cobalt
    tetrasulfophthalocyanine,
    121–122
  dichalcogenides, 80
  dinitro compounds, 88–89
  diorganodichalcogenides, 80
  disulfides, 73–74
  graphene (graphite) oxides,
    109–115
  heteroaromatic N-oxides, 91
  hydrazo compounds, 87
  iodate, 134
  iodine, 38, 134–135
  iron
    tetrasulfophthalocyanine
    and its μ-nitrido- and
    μ-oxodimers, 121–122
  ketones, 84–86

lysine residues of glutamine
synthetase, 65
methyl viologen, 93–94
metmyoglobin, 121
Ni^{2+} 125
nitrite, 121
nitro compounds, 87
o-nitroazobenzenes, 88
γ-nitroketones, 90
o-nitrophenylazo dyes, 88
organylselenium dichlorides,
80
organylselenoxides, 80
organyltellurium
dichlorides, 80
organyltellutoxides, 80
organofluorine compounds,
54, 57, 59
2,4-pentanedione, 90
vinyl compounds, 66
sulfilimines, 73–74
sulfoxides, 74
tellurium, 80
redox isomers, 29
selected bond distances, 29
solubility, 37
structure, 28–30
synthesis, 11–13
tautomerization in aqueous
solutions, 31, 133
thermal decomposition, 37
toxicity, 38
usage
as bleaching and dyeing
agent, 142–143, 146
as coreactant of luminol
chemiluminescence
for determination of
disulfides, 157
iridium, 157
rhodium, 157
for preparation of Ni, Co
and Cu nanoparticles,
125
for preparation of
technetium labeled
radiopharmaceuticals,
125–126
in grafting of
2-methyl-5-vinylpyridine

onto oxidized celluloses,
153–154
in modification of polyorganylsilsesquioxanes,
155
in modification of bitumen,
157
in photography, 157
in polymerization of
acrylates and vinyl
chloride, 154–155
in syntheses of metal
sulfides, 125
zwitterionic structure, 29–30
Thiourea-iodate-sulfite flow system,
135
Thiourea monoxide
(aminoiminomethanesulfenic acid,
formamidinesulfenic acid)
formation, 12, 17
selected bond distances, 29–30
Thiourea trioxide
(aminoiminomethanesulfonic acid,
formamidinesulfonic acid)
decomposition in aqueous
solutions, 44
reactions with
chitosan, 65–66
iodine, 38, 134–135
lysine residues of glutamine
synthetase, 65
methyl anthranilates, 62
2,4-pentanedione, 90
primary aliphatic amines,
62
sodium azide, 90–91
selected bond distances, 29
thermal decomposition, 38
usage in
bioactive guanidines
synthesis, 62, 64–66
guanidine synthesis, 60–66
Thrombin inhibitor Du P714
synthesis, 77
Trialkylthiourea trioxides
synthesis, 14
Trifluoromethylation, 56–57

Trimethylthiourea
    reactions with
        bromate, 14, 136
        chlorite, 132

U

Urea, 12, 44

V

Vat dyes
    reduction, 139–140

Vinyl chloride
    polymerization, 150–152, 154
Viologens
    as electron transfer catalysts, 73,
        82, 92, 93, 151–152, 154–155

Printed in the United States
By Bookmasters